CLASS, CULTURE AND COMMUNITY

CLASS, CULTURE AND COMMUNITY
A Biographical Study of Social Change in Mining

Bill Williamson

Department of Sociology
University of Durham

Routledge & Kegan Paul
London, Boston and Henley

First published in 1982
by Routledge & Kegan Paul Ltd
39 Store Street, London WC1E 7DD,
9 Park Street, Boston, Mass. 02108, USA and
Broadway House, Newtown Road,
Henley-on-Thames, Oxon RG9 1EN
Printed in Great Britain by
T.J. Press (Padstow) Ltd

Library of Congress Cataloging in Publication Data

Williamson, Bill, 1944–

Class, culture, and community.
Bibliography: p.
Includes index.
1. Coal miners – England – Northumberland – History.
I. Title.

HD8039.M62G794   306'.3   81-20971

ISBN 0-7100-0991-7   AACR2

'See to your needs first, then your wants.'
James Brown of Throckley

CONTENTS

ILLUSTRATIONS

DIAGRAM

MAP

PREFACE

I have incurred many debts in writing this book and most of them are of the sort that can never be repaid. Many people have given me their time and willingly allowed me to question them closely. I only hope that if the people I have spoken to read this book they will feel that for all its faults it conveys accurately something of the life of the community in which they spent a good part of their lives.

To name but a few of those who helped is invidious. But I must thank Jack Armstrong, Throckley's local historian, for guidance and access to his photographs, Mrs Donaldson of Heddon-on-the-Wall for having the foresight to keep her father's union records, and the Headmistress of Heddon School for allowing me to use the school log books. Hubert Laws of Houghton Farm, Heddon, showed me photographs and documents; but better still, he has allowed me to explore thoroughly the farm on which my grandfather spent so much of his youth.

Jack Davison of Pegswood, the man who wrote 'Northumberland Miners' History 1919-1939', gave me some helpful clues at the outset of my work. The Vicar of Newburn gave me access to old parish magazines. Mr Walton of Newburn Public Library helped with photographs, and information about Methodism. Albert Matthewson, Tom Stobbart, Charlton and Jean Thompson, all from Throckley, gave me much of their valuable time. John Stephenson of Wylam, former coal owner, helped a lot with the history of the coal company. Mrs Gibb, of Cambo, former northern region organiser of the Labour Party, gave me plenty to think about concerning the Labour Party in Throckley.

The staff of the Northumberland Record Office and of the Tyne-Wear Archive have been incredibly helpful. I would like to thank them for the enthusiastic and impressively competent way in which they have dealt with my queries. I should like to thank, too, the staff of the Northumberland Miners' Assocation at Burt Hall, Newcastle. They gave me access to their records and made me feel very welcome every time I went to use them. Bill Dowding of the Durham Miners' Association at Redhill, Durham, helped, too, with information and support.

My colleagues in the University of Durham, Richard Brown in particular, have steadied my hand throughout, persuading me when I doubted it that the enterprise of writing this book was worth its risks. David Chaney and Robin Williams have helped me with many questions of method and of approach in what came to be known as my 'grandad project'. Martin Daunton, of University

College, London University and formerly of Durham, helped a lot with sources and historical methods.

The main thanks, however, must go to my own family, to my mother, aunt Eva, uncles Bill and Jim and aunt Nelly. I pestered them with visits and telephone calls and they gave me material in abundance. Gloria, aunt Mary, Olive, Francy and Sadie have had their memories dredged, too. The composite picture of my grandparents which has been built up from these discussions is, of course, my own. Without them, however, I simply could not have done it.

The colleague who told me that writing a biography of a working man who had not committed his thoughts to paper was impossible gave me a special if unintended spur. If what I have done encourages just a few ordinary folk to treat their family history seriously, then I shall be well pleased; we might then make more sense of history itself.

Diane, Johnny and Joanna managed to live with me while I was writing. They have checked tables, read proofs and, Diane particularly, searched archives and libraries. They, too, must be relieved that the 'grandad project' is finished. Finally, I want to thank once again Trudy Harrison. There can't be many mothers-in-law who will turn a tatty manuscript into a smart typescript with such ease and skill and still remain friends with the author.

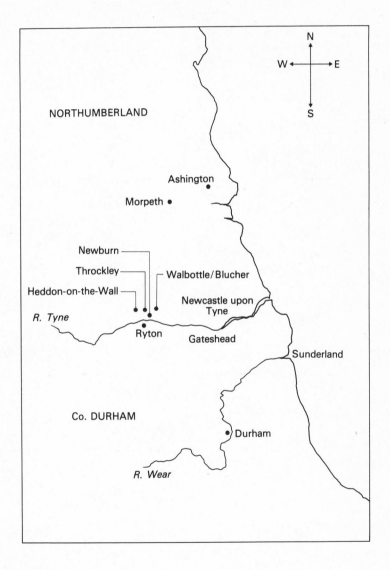

The setting

INTRODUCTION

This book is about the life of my grandfather, a Northumberland miner. Although its focus is on one man and his family it is intended as an account of the changes which took place in the society and community in which he lived. It is written as a biographical study of social change in mining – to be, I hope, a small contribution to the two disciplines I have drawn on to write it, sociology and history.

These themes which dominate these pages are those of class, culture and community, and a principal aim of the book, apart from describing something of the life of the man who is its subject, is to show that biography is a form of writing and analysis appropriate to the study of social change and representing a way of reconciling the work of historians and sociologists. This is, of course, hardly new. C. Wright Mills (1970, p. 159) emphasised twenty years ago that the 'co-ordinate points in the proper study of man' were the problems 'of biography, of history and of their intersections within social structures'. And his point with respect to history and social science, namely that 'history is the shank of social study', is one which I accept fully.

JAMES BROWN OF THROCKLEY

James Brown was, for the whole of his working life, a miner. He was born in a small village in the Tyne valley in 1872 and he died at the age of 93. From the age of 11 onwards he worked in three pits, the Heddon Margaret, the Throckley Isabella and the Throckley Maria, all of them within the same area of the Northumberland coalfield, all owned by the same coal company. Heddon-on-the-Wall, the village in which he grew up, was a small, almost rural village. Throckley, the village to which he moved when he married in 1900, was entirely a mining village. He had seven children. One died within the first year; the rest, of whom four still live, survived to bring up their own families to maturity. Most of their family life was spent in a colliery house, 177 Mount Pleasant, and there is little in that life which distinguishes any of them from everybody else who lived in Throckley. They were an ordinary family and James Brown just an ordinary pitman.

He liked a drink. He was a family man. He looked after his gardens. He travelled hardly at all. After the First World War he voted nothing else but Labour. He spoke in a soft Northumbrian accent with rolling 'r's' and used the archaic pronouns 'thee' and 'thou'. He never tried to leave the pits or the area. He was a hard-working man, well respected in the village. He was uncharacteristically tall for a miner, very strong and not easily roused to anger. He was a member of the Northumberland Miners' Association (NMA) till he left the pits in 1935. And in old age, prompted to talk about his past, he talked of progress and improvement. He died content in his own home surrounded by his family.

These, in obituary form, are the essential facts of his life. Unravelled and discussed historically and fleshed with as much rich detail as the record now allows us to reconstruct, they become illustrative of some of the major, complex moments of social change which have transformed British society from the last quarter of Victoria's reign to the second half of the twentieth century.

MOMENTS OF SOCIAL CHANGE

Social change is pervasive and in a sense total; it embraces everything in society and culture and for this reason it is difficult to describe since any description involves selection and simplification. And no description is free of theorising or avoids making assumptions about what counts as valid explanation in history. There is no space in a book which is more descriptive than analytical in its basic aims to treat any of these philosophical questions fully. I shall, however, make clear what my assumptions are with respect to such issues as understanding and explanation and return to them again at the end of the book.

For the present it is only necessary to note that this book concerns itself with several interlocking trajectories of social change and two major historical transitions. The trajectories of change include the rise of the organised Labour movement and the decline of Liberalism, shifts in the character of mining trade unionism and structural changes in the mining industry itself. In the community I have written about, there is the growth of local institutions, the co-operative store, the chapels, the working men's club and so on. And there is, too, the more subtle long-term transformation in the meaning and significance of community for the people who still live in Throckley. In the family I deal with, there are unfolding patterns of growing up and of new generations being formed.

My grandfather experienced all of these changes directly; they were distilled for him in a continuous flow of events and experiences, and the task I set myself in writing his biography was to show how he ordered those experiences into a coherent view of a changing world and his own place in it.

I describe the historical transitions I am concerned with as the shifts from paternalistic capitalism to corporate capitalism

and the welfare state. My grandfather grew to manhood in a
village which was paternalistic and squirearchical and in a
society which was buoyantly imperialistic and rooted in the
social framework of liberal capitalism. He brought his family up
in a village which, at least until the early 1930s, was dominated
by the paternalism of a single coal company. After the Second
World War he lived out his retirement in a society of corporate
capitalism and the welfare state.

These, at least, are the terms I use to describe, as I see them,
the changes which have taken place in the structure of British
society since the last quarter of the nineteenth century. The
two forms of society are defined by the way in which particular
configurations of production, power, class and crisis relate to
one another. The shift from one to the other is not the outcome
of an unfolding unilinear pattern; it is the product of the
actions of different social groups pressing their inconsistent
claims for their share in the rewards (both material and symbolic)
of industrial production and political power.

It would take another book to explain properly this view of
social structure and for my present purpose it is not strictly
necessary to do so. What I mean by the phrase, 'configurations
of production, power, class and crisis' is this. In the different
time periods covered here, economic production changed in its
scale and organisation. The large corporation employing thousands of people replaced the small firm. State involvement in
the economy, itself a response to crises of different kinds,
combined with both changing social expectations and economic
organisation to define new patterns of class relationships.
Economic and political change transform the distribution of
power throughout society and together such changes modify the
market position, life chances and perceived interests of whole
groups of people, workers and employers alike. At each point
in time the limits and possibilities of social change are circumscribed, if not fully understood, and social change itself is
forced by crisis.

The crises I am principally concerned with are those of the
two world wars, the industrial troubles of the 1920s and the
Depression. In a sense, of course, these events encompass the
whole history of the twentieth century. What I show is that they
had specific meanings for the changing patterns of power in
British society, and that it is these specific meanings which
fuelled demands for change which could only be met by older
social structures giving way to new forms of social relations.
The most visible and well-documented shift has been the growing involvement of the state in the regulation of economic life
and, within that, the growing involvement of the organised
labour in the affairs of the state itself.

The task I set myself when I began this work was to see in
what ways the changes I have described penetrated the village
in which my grandfather lived, and shaped the experience of
men like him.

QUESTIONS OF APPROACH

My work has its roots in two largely distinct traditions of
inquiry. The first of these is a tradition of sociology extending
back to the work of Dilthey and Weber in Germany and to Mead
in the USA. The guiding propositions here are, firstly, that
the actions of men are subjectively meaningful and the explanation of their action involves taking into account the viewpoint
of men themselves; secondly, that the subjective world of the
person is a shared one patterned by culture.

Acknowledging these propositions, I have been concerned to
so portray the life of my grandfather and the structures of
community in the village in which he lived that the meanings he
attached to his actions and which he shared subjectively with
significant others in the village can be made clear.

The second tradition of inquiry is a growing body of research
and writing in the field of labour history concerned to portray
faithfully something of the lives of those whom Lucien Febvre
once described as 'doomed to do the donkey work of history'
(Febvre, 1973, p. 2). There are many autobiographies of working people available (see Burnett, 1977), and the oral historians
have extended our awareness of the richness of recollection and
the 'depth' of oral traditions (see Evans, 1976; Thompson, 1978;
Meacham, 1977). But ordinary people do not appear through the
mists of time simply as writers or raconteurs; they exist, too,
as poets and preachers, singers and villains, unsung activists
in politics and trade unionism (see Colls, 1977). In the receding
memories of those who have succeeded them they appear as
fathers, uncles, grandparents and friends. The kaleidoscope of
their experience has always been a rich seam for novelists. My
hope is that the re-creation of that experience, or its redis-
covery, should not be something left entirely to the imaginative
pen of the writer; it is meat, too, for the historian and the
sociologist.

The character of the work in which I have been engaged bears
comparison with that of, for example, Stella Davies (1963),
Robert Roberts (1971, 1976) and M.K. Ashby (1961). These
books take the form of family histories and each has a strong
autobiographical purpose. Davies deals with the family history
of both her husband and herself, tracing them back to the
sixteenth century. The main period covered is, however, that
of the nineteenth-century industrial revolution and the first
half of the twentieth century. The book itself is a vivid and
kaleidoscopic picture of the birth and transformation of the
Lancashire working class.

In Roberts's books he brilliantly reconstructs the life of a
Salford slum in the years before the First World War. They could
only have been written by an 'insider' and then only by one as
sensitive and observant as Roberts himself. The same can be
said of Ashby's book which deals with the life of her father in
Tysoe, a small rural village in Warwickshire. She is too self-

effacing to describe her work as history on the grounds that
'it relies too largely upon memories and oral tradition - on family
and village stories, reminiscences of table-talk, of daily life,
of speeches heard and occasions shared' (Ashby, 1961, p. ix).
But history is what it is and the book is richly evocative of
nineteenth-century rural England.

All these books, in fact, succeed in communicating a sense of
change mediated directly in the experience of individuals and
their families; they succeed because they personalise social
change. My aim, with the same unashamed focus on one man and
his family, my own family, is to record something of the experience of social change in a mining community.

CLASS AND COMMUNITY

The analytical purpose of my work has been to understand more
fully the potent force of class in British society and to clarify
what is meant by the idea of community. Briefly stated, my
position is this. Class is not a category; it is a relationship
among men and it is rooted ultimately in the organisation of
economic life and the social relationships of production (cf.
Thompson, 1968). Class analysis, as I understand it, is essentially a tool of history since what is of central importance to it
is the way class relationships change through time.

The focus of this book on two small communities, and on
Throckley in particular, raises many questions about the value
of community studies and about the appropriateness of different
methods of studying community history. In recent years community studies have come in for a great deal of criticism. It has
been argued that they have contributed little to general theory
in the social sciences (Elias, 1974). A. Macfarlane (1977) has
pressed the point that the use of anthropological methods in
community studies has tended to cause a neglect of history. The
result, he feels, is that many such studies portray a false picture of the social relationships in small localities, invariably
stressing features such as integration and social cohesion to the
exclusion of conflicts, change and instability. Margaret Stacey
(1969) has argued that the notion of community is very impre-
cise; it is not clear whether it is defined in terms of geographical
space or some vague sense of belonging. She suggests that
instead of studying communities sociologists should study how
social institutions interrelate in particular localities. I have made
no attempt to adjudicate in these arguments; I have tried, however, to remain aware of them.

My point about community is threefold. Firstly, and here I
agree with Stacey, studies of particular localities must recognise
that localities are not in some way isolated. What I show in the
cases of Heddon-on-the-Wall and Throckley is that while much
of the life of those villages must be understood, as it were, 'on
its own terms', their social structure was none the less

massively shaped by society as a whole.

Secondly, the notion of community embraces not just the idea of locality or social networks of particular kinds; it refers to the rich mosaic of subjective meanings which people attach - or in the special case of this study, attached - to the place itself and to the social relationships of which they are a part (cf. Thorpe, 1970). It is in terms of such meanings that the community can be recognised and the people who live there can recognise themselves. The pattern of these meanings is what constitutes the culture of the community.

The geographical location of the villages is only important to my account in so far as people themselves placed some importance on it. One of my themes, in fact, is that the mining community of Throckley disappeared in the period from the late 1930s onwards. Throckley is still where it always was; evidence of mining is there for all to see; old people still recall the mining community. But in so far as it exists now it exists as images in fading memories.

The disappearance of the mining community leads to my third and perhaps most important point. Throckley did not always exist as a mining community; it was built up as a community of miners quite deliberately by the coal company which sank the pit there in the late 1860s. I describe it as a *constructed community*. Some of the qualities which have been associated with mining communities can certainly be traced in Throckley. Geographical isolation, traditionalism, a suspicion of strangers, great solidarity among the men and a clear sense of Us and Them - features which, C. Kerr and A. Siegel (1954) once argued, explained the militancy of miners - can all be found. Tight bonds of kinship, the clear separation of the roles of men and women, and occupational homogeneity, i.e. some of the key characteristics of the ideal typical mining community described by Martin Bulmer (1975), are also in evidence.

However, such qualities have to be grasped historically. They have to be seen as part of a moment of historical change when the special circumstances of capital investment in mining require the creation not just of labour camps, but of communities. They have to be understood, too, against the uniqueness of particular villages, for while mining communities had much in common there were, none the less, subtle but powerful differences of structure, experience and attitude which associated with very different political and industrial behaviour (Harrison, 1978).

The same moment of change spawned a response among miners themselves. Through their unions and co-operative societies they built their own institutions distinct from those of the coal company. Through family and kinship they built defensive walls against chance and circumstance, constructing a way of life which was theirs and not simply a reflection of the coal company's plans.

This is, of course, the point where my focus on community connects with my concerns about class. As I use the term,

class relationships refer to relationships among groups of men.
The form of those relationships determines the social distribution
of opportunities and life chances with respect to work, income,
housing and personal development. Class relationships are by
definition relationships of superiority and subordination reflecting the distribution of power in society.

But class is also something which is experienced; it is a mode
of social recognition bringing, under certain conditions, a consciousness of belonging. From this viewpoint class is like a
mirror. In its reflection a man can recognise himself and others.
And the recognition is instant. Small clues – talk, dress, accent,
gait – are all that is needed to recognise a much larger pattern.
Us and Them can be sharply defined. Common experience,
shared anxieties and hopes are at the base of it. The boundaries
of that recognition, that sense of belonging, are intensely local
(see Williams, 1977). The family, the street, the village and, in
this book, the pit, define most of them.

But these are permeable boundaries. Village life, lived in the
present tense, often involving rivalry with other villages, is
transcended, though never with any certainty, by a collective
recognition extending hesitantly to a feeling of common cause
with other working men, perhaps, even, to the abstraction of a
working class as a whole.

Such a shift is not a natural one; nor is it ever-present. In
the history of the labour movement in Britain it was something
which had to be forged by politics and it has always been contingent and precarious. The kind of recognition which Karl Marx
described referring to classes 'in themselves' has a firm base in
the social encounters of everyday life. But the recognition
implied by social classes 'for themselves' is altogether of a different order. Indeed, much militates against it. Dominant ideo-
logies blurring distinctions between capitalists and labourers
are only part of the problem. Social divisions among working
people themselves are equally important as corrosives of a
theoretically expected class consciousness.

My aim, however, is not aridly to construct a theoretical
scheme of linked concepts standing aloof from history. I simply
want to set out some of the elementary terms I have found to
be helpful in making sense of the actions and experience of my
grandfather and changes in the political life of the village I
have studied.

My conception of the village, viewed statically, is represented
in the diagram on page 8. I have tried throughout this book to
connect the structures of the village to the wider society in
which they were found and to trace in the specific relationships
between the Throckley coal company and the men it employed
the pattern of opportunities and experiences which the generic
term, class, summarises. The coal company appears in these
pages as the representative of industrial capital, of ownership
and of the political dominance of a bourgeois class.

The Throckley miners' lodges, the Labour Party, the co-

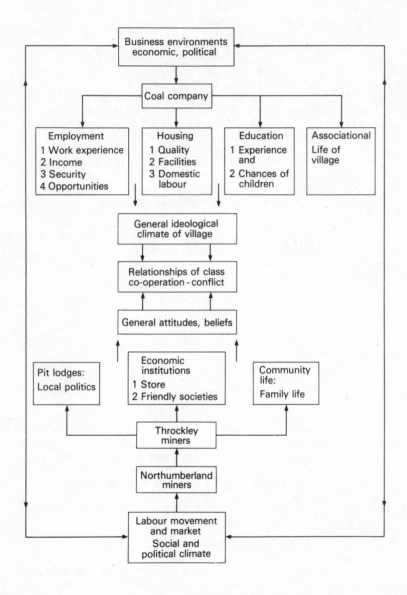

Social structure in a mining village: a model

operative store symbolise organised labour. The relationship
between these two forces defines the character of Throckley as
a community. Changes in the community can then be interpreted
as elements of the moments of social change itself which I described earlier.

The diagram does no more than indicate how I have constructed my account of the village and its principal institutions. The
account itself takes seriously, however, that these institutions
existed in time and that their significance in the life of the
village changed through time.

This is particularly true of social conflict. As in all mining
villages, the division between owners and workmen was absolutely clear. While the potentiality for conflict between the two
groups was always there - part of the organisation of pit production itself, as I show - there were long periods of industrial
peace and short periods of intensive conflict during the years
covered in this book. Most of the time Throckley pits were
peaceful; it was not a militant village. Until the late 1920s the
coal company was considered a good one to work for and the
owners were well respected. Larger conflicts of organised labour
and capital found their expression elsewhere in the growth of
labour politics, in the co-operative movement, in political processes at county and national levels. The community itself,
then, did not encapsulate all the structures of class which shaped
the opportunities and experiences of Throckley pitmen. But it
was, none the less, an integral part of that experience and their
sense of who they were.

The experience of class itself, the subjective feel of it, connects closely with the theme of biography. Biography is a form
of understanding which in principle could illuminate well what
class feels like, what it is to be a member of a particular group
in society; for class feelings, as Sennet and Cobb (1972) have
brilliantly shown, involve the self. They shape a person's sense
of his personal worth for they follow socially sanctioned ways
of recognising inferiority and superiority. These subjective
feelings, at least with respect to those thought to be inferior,
are what Sennet and Cobb call 'the hidden injuries of class'.

A recurring theme in my account of my grandparents concerns
the way in which they, and others like them, parried the
injuries of inferiority which defined their social position by
building up a basis for their own self-respect in ways which
neither the coal company nor economic circumstances could
affect. They did not just assert they were as good as anyone
else. Against much in their experience, particularly their
education, which persuaded them of their inferiority, they built
up, over the years, a reputation in work, in the home and in
their dealings with neighbours, which gave them dignity and
respectability as people, quite apart from their limited social
roles as housewife and pitman. Probing closely, with sensitivity,
the private world of the person - and this I take to be the hall-
mark of biography - is one way of focusing on this dimension

of class.

But class shaped them in a way directly related to this biography. They conceived of themselves very much in the terms
that society conceived of them – namely as just ordinary people.
Being like everybody else they were never disposed to see their
own lives, particularly their own history, as being of interest
to anyone. Indeed, I am absolutely convinced that if my grandfather could know that I had written about him he would consider it a monumental waste of time. Not placing any public
value on his opinions, he kept them very much to himself. He
can be respected for that, but it has made the task of describing
those opinions a lot more difficult. His reticence and lack of
interest in himself are, however, subtle injuries of class.

THE PERSON AND BIOGRAPHY AS METHOD

The person appears in sociology as an actor or player of roles,
a prisoner of circumstance or an agent of change. Rarely, in
contrast to psychology and to psycho-analysis in particular, is
the person conceived of as having a past, as someone who exists
in time. And until recently the person has simply been ignored
as a focus of analysis. The reasons for this are connected with
the growth of a scientific attitude in social science in which the
logic of large numbers, of generalisation and empirical research
methods, has displaced that of the interpretive understanding
of social action.

This has not always been the case. Wilhelm Dilthey, the
German philosopher and one of the fathers of interpretive
sociology, gave the study of individuals a central place in his
historical methodology. 'For him', writes H.P. Rickman, (1976,
p. 35), 'the biographies of historically influential men are the
natural building blocks of history because the pattern of a man's
life provides a principle of organisation of diverse themes.'
The thinking individual is the fulcrum of our understanding of
the interconnectedness of a whole culture. For this reason
Dilthey saw autobiography as 'the highest and most instructive
form of the understanding of life' and the study of how individual men reflect on their own lives is what 'alone makes historical insight possible'; such reflections, he says, 'are the
foundations of historical vision which enables us to give new
life to the bloodless shadows of the past' (ibid, pp. 214-15).
For these reasons Dilthey's work is rich in biographical sketches
of particular individuals.

In similar vein W.I. Thomas and F. Znaniecki argued in their
classic study, 'The Polish Peasant', that social institutions cannot be fully understood unless studied in relation to the personal
experience of their members. For this reason they suggest that
'personal life records, as complete as possible, constitute the
perfect type of sociological material' (1958, p. 1832). More
recently Howard Becker, discussing how important life-history

studies were to the Chicago school of sociologists, has argued
that such techniques, focusing on single individuals, are a
'touchstone to evaluate theories' giving insight into the subjective side of institutional processes, 'a live and vibrant message
from "down there" telling us what it means to be a kind of
person we have never met face to face' (Becker, 1971, p. 70).

Life histories and biographies are not the same thing, the
former being largely under the control of the person or subject,
whereas biography is written by someone else. For the moment,
however, such a distinction is unimportant; I merely want to
stress that both are concerned with the subjective experience of
individuals, and to be of any value at all they must locate that
experience in time.

Biography has an external and an internal aspect. Externally,
it relates the way in which individuals move in the course of
their lives through the roles their society lays out for them to
play. Internally, it deals with the socially mediated significance
of those roles (Mills, 1970; Gerth and Mills, 1954). The roles
men have played shape powerfully how they act, think and feel;
they are a major component of a man's self-image and self-
respect.

But biography is something which men act on, too; it is under
constant reinterpretation, and through this process of self-
understanding men come to understand the course of their lives
in particular ways. This process of understanding is an integral
part of their consciousness of themselves and of their society.

In addition to the two basic assumptions that the individual
has a history and that people think about themselves and live in
a meaningful world, I make a third. It is that the life of an
individual is embedded in a complex of everyday routines which
clothe it with a taken-for-grantedness and matter-of-factness
from which it is difficult to stand back. The world of everyday
life is a highly structured world; it communicates to the individual a sense of order and makes available to people conventional
ways of making sense of events and experience (Berger and
Luckmann, 1966). At the same time the individual acts on that
world; he must, as Thomas and Znaniecki pointed out, 'consciously define every situation', constructing in a reflective way
general schemes of understanding of self and society which they
call 'life-organisation' (1958, p. 1852). In this light, the kinds
of men or of situations are interesting which effect change in
patterns of life organisation.

Central to my notion of life organisation is the notion of commitment (cf. Becker, 1970). This refers to the decisions people
make about many different aspects of their lives, e.g. their
work, family life, trade union activity, leisure and so on, which
reflect their priorities and their sense of obligation and which
result in consistent forms of behaviour over time. Commitments
are interdependent and therefore have implications for each
other. As Becker has analysed the term, commitments involve
'side-bets' so that consistent behaviour in one sphere of

activity, e.g. work, is likely to imply consistency in another
sphere such as family life, and the one can reinforce the other
making it extremely difficult for people to make radical changes
in the pattern of their lives without incurring great social and
psychological costs.

The character of particular commitments is shaped by prevailing social values and expectations. Commitments themselves,
given this particular usage, are not necessarily the outcome of
explicitly conscious decision making. But they do refer to what
people think of as important in their lives, what is worth working for, what is worth defending and worth worrying about. I
show in this book that, given the specific range of opportunities
in Throckley, my grandfather gradually evolved a series of
commitments - many of them different from those of other men
in the village - involving his work, his family and his gardens
which reflected his priorities, distinguished him from others
and gave pattern and meaning to his behaviour over many long
years, and that these commitments were woven into his everyday routines.

I try to show that for much of his life my grandfather never
stood back from the routines of his everyday life to question
them. In this respect he was an unreflective man. Indeed, my
argument is stronger than that. There was little around him to
encourage him to do so and much to prevent it. His education,
the character of the work he did, the stance of his union leaders
and local politicians were all part of a broader ideological framework or climate in which men like him were persuaded, although
never completely, to accept their everyday life and conditions
as something normal and inevitable.

But there were times when he did reflect; there were great
formative moments of change when his understanding of himself
and his world changed. These were the moments of crisis. They
include the First World War and the miners' lock-out of 1926;
but they also include events much closer to him in space and
time such as accidents in the pit, the death of a friend or
events in his family life. And how he modified his understanding
is, in essence, the process which we call social change, the
intangible metamorphosis against which the more obvious changes
of law, policy and power in society as a whole must be seen.

Such changes, I hope to show, are amenable to study using
biographical techniques and methods of life-history research.
There are, however, some serious methodological problems to be
faced, particularly if the man being studied did not express
himself in print.

PROBLEMS OF METHOD

Life-history techniques of research, biographical studies and
other kinds of materials using 'personal documents' have fallen
out of favour on the grounds of their scientific inadequacy as

evidence. H. Blumer has raised pertinent questions about the adequacy, reliability, representativeness of such data and of the validity of the ways in which they are interpreted. Martin Bulmer has argued, however, that Blumer's criteria, though valid in themselves, do not amount to rejecting the use of personal documents related to the life of particular individuals. Used in conjunction with other kinds of data, a method known as 'triangulation', such accounts serve as a rich source of material in their own right and as a basis of 're-analysis and fresh theoretical interpretation' (Bulmer, 1978, p. 309).

H. Schwartz and J. Jacobs (1979) have recently endorsed this view, claiming that life histories of single individuals can be representative of a larger group of people and 'an independent totality' from which generalisations may be drawn. It remains true, however, that those who are interested in life-history techniques, biography and the unique individual are also those committed to a particular tradition of interpretive social theory (see Faraday and Plummer, 1979). Also, little agreement exists about the appropriate methods for studying life histories or about the limitations of the materials which can be collected in the overall programme of social research.

The methods I have used in writing about my grandfather exploit these approaches although I do not rely on them exclusively. I have spent a lot of time talking to my relatives about their past. My sense of what the important themes of the biography had to be was largely built up from those conversations. I have not taken the view, however, that such accounts, as it were, speak for themselves. I disagree with Schwartz and Jacobs, for example, who, emphasising the importance of writing so-called life histories as a method of social research, insist that the golden rule of such work is: 'Believe what you're told' (1979, p. 72). Like M.K. Ashby, I, too, feel that the data of reported experience have to be set alongside other kinds of data so that they can be read in context and be corroborated.

Checked in detail against conventional documentary sources the value of personal reminiscence is high; 'triangulation' is an ugly word to describe this process but it is a very necessary task.

There are two additional features of my methods which need emphasis. One concerns the place of imagination in historical work and the other the nature of the historical enterprise itself. My contention is that to portray faithfully the experience of people requires imagination and empathy. The techniques available to us as members of society which enable us to take the role of another person, to see the world how others see it, thereby helping us to understand them, are techniques essential to historical and sociological work. The imaginative reconstruction of the behaviour of others is an integral part of our ability to empathise with them and, therefore, to understand them (see Rickman, 1976).

The question is raised, of course, whether the understanding

arrived at in this way is literary rather than historical. My
answer, simply, is that it is both and that it is wrong to drive
too strong a wedge between history or sociology and literature.
Both forms of understanding have techniques in common though
their aims may be different (see Fischer, 1976). History, sociology and literature have as their data or sources various forms
of human expression - documents, rules, letters, stories,
physical artefacts and so on - which carry the essential clues
to the thoughts and feelings of men in the societies they come
from. In the case of literature the materials are, of course,
imaginative; in history the onus is on the historian to deal with
real persons and events of the ascertainable past.

As to technique, narrative is the essential form of both history
and literature (see Mink, 1970). The telling of a story is a
device to convey understanding, the aim, in fact, of both history and literature. Narrative frames events in time; it links
events and experience into a coherent sequence and this, I
maintain, is a form of understanding. The technical work of the
historian or sociologist in establishing the evidence around
which the story is based is what distinguishes historical understanding from that of literature, but the two can be fused in the
genre of the historical novel. Essential to both, however, is
the disciplined use of imagination and empathy and an ability to
relate events, artefacts and experience to their broader con-
texts of society and culture. Historical research differs from the
writing of historical novels, at least from this point of view,
because of the commitment to understand events which did
actually occur in the past.

Such arguments about imagination, understanding and the
essential tasks of history and social science, have been forcefully made by writers in the so-called 'Annales School' of French
history (see Burke, 1973) and, as an attitude to literature,
by John Berger (1969). Without claiming my work to be representative of these approaches, I have none the less felt free to
reconstruct imaginatively those thoughts and feelings of my
grandparents which they never articulated to their own children
and which, but for reconstruction, would be lost entirely. Such
reconstruction is not arbitrary; the artefacts I have worked on
to be able to do it at all are the reports of my grandfather and
his actions which have come to me through my family. I have
set these against documentary records and the accounts of
other people in the village I studied.

This brings me to my second point; it is not directly methodological but it bears on method and concerns how the kind of
interpretation I have been involved in actually gets done. My
account of my grandfather is clearly a subjective one, and with
respect to some issues, e.g. his entry to the pit as a boy, it is
based on a form of imaginative reconstruction of experiences he
might have undergone but about which I have no certain evidence. But it is not entirely a subjective and idiosyncratic
account for at each stage I have tried out my interpretation on

people who knew my grandfather well. They have read what I
have written; I have tried out my arguments on them to see if
they 'strike true'. Together we have interpreted my grandfather's experience. The words used might, in the end, be
entirely my own. But the ideas they seek to convey were developed co-operatively.

The form of work in which I have been engaged, although not,
at least when I first began it, consciously, is close to that cur-
rently being developed by the Ruskin History Workshop in
Oxford. Their aim is to make history 'relevant to ordinary
people' ('History Workshop', no. 1, Spring 1976). Their journal,
'History Workshop' 'is dedicated to making history a more democratic activity and a more urgent concern'. They go on:

> We believe that history is a source of inspiration and under-
> standing, furnishing not only the means of interpreting the
> past but also the best critical vantage point from which to
> view the present. So we believe that history should become
> common property, capable of shaping people's understanding
> of themselves and the society in which they live. ('History
> Workshop', no. 2, 1976)

This view of history not only expresses my own purpose well;
it also, it seems to me, represents an effective way of organising
research and generating the data from which social history has
to be written.

Working this way does not, however, guarantee truth or
prevent distortion for it portrays experience in ways that those
being portrayed would not themselves use. It draws connections,
which people might not themselves make, for reasons of theory
or of analysis which interest the subjects not at all. This is the
distortion of taking an analytical attitude. There is no way that
I could ever become an insider and make sense of their social
reality as my grandparents experienced it. The result is that my
account is an interpretation which might in several major
respects be wrong. I must, however, take the risk. In any
case, history cannot be re-lived. But since history is no one's
particular property others can plunder it just as freely as I
have done to make of it what they can.

This book is organised into thirteen chapters including a
short conclusion. The first three deal with my grandfather's
life as a child, youth and young man up to when he married.
Chapters 4 to 8 describe the period when they moved to live in
Throckley and brought up their own family. Chapter 9 discus-
ses the First World War, the first turning point in my grandfather's life.

The following two chapters discuss the industrial troubles of
the 1920s and 1930s including the General Strike. The theme
for these chapters is industrial defeat and class politics. Chapter
12 examines the changes wrought in Throckley by the Second
World War and returns to the theme of how British society as a

whole had changed.

The short concluding chapter tries to make explicit what the value of my account really is, placing it in a comparative perspective and emphasising that studies of this kind need to be complemented by similarly conceived attempts to interpret the experiences of different villages and generations of working men, in different industries, in different parts of the country. Only in this way can the rich experience of ordinary folk penetrate 'the enormous condescension of posterity', to use E.P. Thompson's striking phrase.

1 HEDDON-ON-THE-WALL AND FRAGMENTS OF A CHILDHOOD

'O World, how apt the poor
are to be proud'

Shakespeare, 'Twelfth Night'

The village in which my grandfather was born and in which he
grew to manhood lies eight miles west of Newcastle high along
the valley of the Tyne and Hadrian's Wall. Today it is an estate-
rimmed commuter village, a dormitory for professional middle-
class people who work in Newcastle. There remains a nucleus of
the old Victorian village. There is a village green, an ancient
church, St Andrew's, a Wesleyan chapel built in 1877, an early
Victorian vicarage, slightly shabby and certainly not as proud
as it once was, a school and the remains of a busy blacksmiths'.
The smithy is now a pub in late twentieth-century old-world
style attracting the motoring public with keg beer and a softly
lit restaurant. The reading room is being converted into an
executive residence, and the few people who knew Heddon as it
was, before the First World War or the turn of the century,
are fast fading away.

The village is described by estate agents as attractive. To its
south lies the river, curving into Tyneside in one direction and
into the hills in the other. To the north, on clear days, the
Cheviot hills can be seen standing proud on the horizon. The
village is surrounded by neat farms and on its western border
by the carefully laid grounds of Close House, the former home
of the Bewicke family and now owned by the University of
Newcastle-upon-Tyne. The pit, brickworks and quarry which,
apart from the land, gave the citizens of Heddon their work
have long gone. The village, once semi-rural, is now fully
absorbed into the life of the conurbation as a whole; most
people who live there do not work there. It is a village of expensive houses known throughout Tyneside as a desirable place to
live.

In 1872, the year my great-grandparents came to Heddon, the
village was busily organised around farming, quarrying and
mining with a social structure and social life which was distinctly paternalistic. It was a paternalism based firmly on the
ownership of land and the authority of the leading families of
the village was largely unquestioned. There were, of course,
long-term changes in the whole structure of society which were

eroding that authority - the rise of trade unionism in agriculture
and mining, and the emergence of a powerful new class of industrial capitalists represented in Tyneside by such families as the
Cooksons, Ridleys and Spencers and by such men as Lord
Armstrong and Sir Andrew Noble whose wealth was based in
heavy engineering and coal (Benwell Community Project, 1979).
But Heddon was still clearly dominated by owners of land.
John Clayton of Chesters, for example, who owned much of the
land around Heddon, held over 11,000 acres in 1883 with a rental
value of over £13,000 annually (Benwell Community Project,
1979, p. 35). And the Heddon squire, Calverly Bewicke, was
heir to an estate of over 2,500 acres dominated by the early
nineteenth-century mansion of Close House.

In this and the following two chapters I shall describe the
village context in which my grandfather spent his childhood
and something of his family life. I shall describe his schooling,
his early start to work in the pit and a little of his youth. My
aim is to show that his experiences as a boy and a young man
were powerfully shaped by his class position.

Class, as I argued in the Introduction, is a pattern of social
relationships whose regulating principles determine the roles
people play and the opportunities which are available to them.
It is a system of domination, injury, constraint and social recognition which shapes experience, and it is confronted first in
childhood. Through my account of the position of my grandfather's family in Heddon, his work and his youth, I seek to
illustrate the ways in which structures of class fashioned an
early sense of himself and his worth as a person.

In the rest of the book I show how changes in the structures
of class relationships brought about partly, at least, by working
people themselves altered his perception of himself and the
opportunities available to him and his like. And these alterations,
I hope to show, are the changes against which the rise of the
organised labour movement in this country has to be understood.
They are the profound changes of consciousness which transformed the political culture of this country. But this is to move
too far ahead.

My grandfather's images of his childhood, for reasons I have
already explained, are available now only as fragments, filtered
through his family. But these images of images, checked in
research, fit well into a more precise portrait of the man and
lend force to this claim: much of his adult life can only be understood against the poverty, insecurity and indignity of his days
as a boy in Heddon. He defined his own life at critical moments
against those experiences. Through his work, his initiative and
his family life he sought to build a basis for his own self-respect
which much in his class position both had and still did deny
him. And, because throughout his childhood his life was lived
close to the land and its rhythms, he never became fully absorbed
into the urban industrial culture in which he spent the greater
part of his life. Whether, ultimately, he was successful is

something others can judge; what remains true is that recalled
images of his youth formed a benchmark against which he
measured his own achievements and the course of social progress itself.

HIS PARENTS

I must begin with his parents. Little is now known about them,
but it was they who communicated to their children distinctive
attitudes and values which figure strongly in my account of my
grandfather. The most powerful of those were self-reliance and
independence. They avoided demeaning contact with their
superiors and they put their trust not in charity or fate but in
their own wit and hard work.

His parents came to Heddon from Norfolk. She referred to
her husband always as 'Brown'. They had met when the travelling fair, in which he was temporarily employed as a groom for
the horses, passed through King's Lynn. Married at 16, they
already had two children when, at the age of 18, they migrated
north. They travelled by boat from Norfolk to the Tyne and
settled in Heddon-on-the-Wall.

Rural migration in the Victorian age was fuelled by a search
for better wages and the social attractions of expanding towns.
What the specific reasons were in this case is not known. Perhaps they left to escape work in a labour gang tramping from
one place to another, contracted out to farmers, living under
the harsh discipline of a gang-master. Norfolk at this time, the
1860s and early 1870s, was notorious for the depressed state of
its landworkers and the low wages. And at the time they left
they could not have looked with much confidence to a future
where things would be different. The depression in agriculture
was making the lives of labourers even more difficult and rup-
turing the social ties of villages and creating a bleak outlook.
They may even have considered emigration to Canada or New
Zealand and might have received help from colonial governments
in doing so. Perhaps they sought better housing. Rural housing
in the 1860s was grossly overcrowded. But the reasons they
themselves would have given are simply not known.

What probably attracted them most was coal. Between 1854 and
1914 the annual output of the 'Great Northern Coalfield' increased
from 15 to 56 million tons and the labour force increased from
50,000 to 200,000 (Rowe, 1973, p. 8). The growth of coal fuelled
manufacturing development in the region and fed a flourishing
export market. My great-grandfather, known as Norfolk John,
brought no mining skills with him although he was clearly willing
to learn for, judging by his subsequent career, he was prepared
to try his hand at almost anything. At various times he was a
miner, a general labourer in a brickworks, a carter and a part-
time farmer. His young wife was a competent housewife, and had
in addition an interest in carving jewellery, including brooches,

from the jet she collected on the beaches of north Norfolk. It is
said that she carved jet beautifully. However, it was not a skill
which she could develop in Heddon; family life and work consumed all her time.

In his old age Norfolk John used to refer back to a time in
his youth when he worked on a treadmill, and throughout her
life his wife feared the prospect of the workhouse. The treadmill was either part of an irrigation system or was in a reform-
atory; no one in the family knows, although some of his grandchildren still have a chilling image of it. The workhouse fear
was real enough. My great-grandmother was born in one.

I mention this for these early experiences both set a mood
and form a context within which the attitudes of their children
were formed, all of them being encouraged to achieve, through
work, an independence from charity and authority which had
been absent in their own young lives.

These, facts too, reinforce the view that their move north was
an escape from painful experience. They were, therefore, able
to communicate directly and vividly to their children, still in
the age of the workhouse, a picture of rural poverty extending
back to the mid-nineteenth century. These were images my
grandfather retained, forming his sense of where he came from
and what he would do. And they illustrate a more general point:
experiences which shape us go much further back in time than
our own biographies, and the potent principle of class or
position is grasped early in a person's life.

HEDDON-ON-THE-WALL AND THE PATERNALISM OF SQUIREARCHY

Why they should have descended on Heddon is a mystery, but
work was available at Heddon pit, and housing, too, by the side
of Heddon Common. If it had been by chance then they might
have considered themselves fortunate. The village was well
served and well placed for work in the pits and on the land.
There was a school and a public house and a small but growing
community of about 400 people who, among themselves, supplied
most of a couple's main wants. It also offered opportunities for
enterprise. The countryside around was a source of food and
the village lacked transport. Between spells of work at the pit
my great-grandfather ran a small carrying business from Heddon
to Newcastle and Wylam, and the horse could graze freely on
common land.

On this same common land they could and did keep pigs and
a cow, and although their cottage was small, with only two
bedrooms, it was at least well situated. Quarry Cottage was
rented from the squire at Close House, Calverly Bewicke. It
stood back from the road on a small hillside overlooking rolling
farmland which stretched right to the Cheviots. They were 200
yards from Law's farm where they could get fresh milk, eggs

and chickens, and the same distance the other way got them
effortlessly to the pub, the Three Tuns, owned in the early
1870s by William Armitage who also ran the blacksmith's shop.

It would not have taken them long, feeling their way into a
new place, to realise that Heddon was a village dominated by
two families - the Bewickes (and, later, Sir James Knott) and
the Bateses of Station Bank, the former being the major landowners in the district and the latter the owners of the mine.
There were other important landowners who saw Heddon as their
village, for example, the Claytons of Humshaugh and Chesters
and the Freemans of Eachwicke. These prominent families were
benefactors to the village subscribing to the school, the village
institute (provided by Mr Clayton) and, most important of all,
to St Andrew's church. They were the major employers of the
district; Mr Bates was the justice of the peace and they all
served, at different times, on the local Board of Guardians.

Through the provision of employment, charity and public
service and drawing on the social values of the landed classes
from which they came and with which they identified, these
families maintained a social order which was distinctly paternalistic. They clearly felt their obligations to the village acutely
and as gentlemen they looked for respect and an acceptance of
their authority. Mrs Hall, an old resident of Heddon, noted,
for example, when interviewed about her life in Heddon before
the First World War, 'if you worked on the farms or for the
gentry, elections times you had to vote what he said. He used
to tell them what to vote. They expected you doing it' (NRO
T/114).

Heddon was, in this way, typical of agricultural villages
throughout England; its leading families drew on a traditional
justification of their authority, and involved themselves closely
with the life of the village.

Howard Newby has argued that through close involvement with
the villages such people were able, by an 'ideological alchemy'
focused on ideals of community, to 'convert the exercise of power
into "service" to those over whom they ruled' and to form 'a
rigid and arbitrarily controlled hierarchy into an "organic"
community of "mutual dependency" in which they exercised their
obligations through assuming the responsibilities of leadership
and through their periodic doles of charity and patronage'
(Newby, 1977, p. 55). Their paternalism presupposed personal
contact and the closer they became to the people - although they
could never get too close, the social insulation between classes
was high - the more secure was their authority and the less
likelihood that those over whom they ruled would question it.
Such an analysis fits the Heddon case well enough. Miss Sarah
Elliot, an old Heddon resident, put it nicely when talking about
the way the Bewickes at Close House indulged their employees.
'The cooks', she said, 'was allowed the dripping' (NRO T/117).

HOME LIFE FOR THE BROWNS

In the subtle structures of hierarchy by which the people of
Heddon measured themselves, the Browns had a lowly position.
They had come as immigrants. In this respect they were little
different from the itinerant labourers of the land. Their thick
Norfolk accent marked them off as strangers and they were, of
course, poor.

They had brought two children to Heddon from Norfolk,
Jenny and Tom. My grandfather was born in Heddon shortly
after they arrived, and in a state of almost permanent pregnancy
my great grandmother gave birth to nineteen children altogether,
of whom sixteen survived to adulthood and some of them, in
fact, to ripe old age.

The children were delivered in the cottage by a local midwife.
Any medical help they needed had to come from Wylam, two miles
up the river, and the doctor came on horseback. It is astonish-
ing that so many of the children survived. The house was
excessively damp. A later occupant of this property told me
that, because it was built into the bankside, when the rain came
it came in so hard it had to be bailed out with buckets. It was
inconvenient, too. There were only two bedrooms. They slept
'top to tail', as my grandfather used to say, and got their
water from a well. The sanitation comprised an outside midden
'netty' or toilet, and 'midden' for rubbish and they had a persistent problem of rats. My grandfather used to say that killing
rats was a regular job for the children. So, too, was breaking
up sandstone to sprinkle a covering on the earthen floor of the
cottage.

Childhood for the older Brown children was swiftly curtailed
by the growing obligations of work; they got little attention
from their parents. Their upbringing was, as my aunt Eva
described it, 'very rough and ready'. Their mother was always
preoccupied with babies and this meant the children were
drafted early into housework. Norfolk John got the children to
help him in the garden. They fed the pigs and milked the cow.

When they grew older they learned to measure the contribution their father really made to the home, finding him a man of
unpredictable effort and temperament. My grandfather grew to
mistrust his father - even, perhaps, to dislike him. He kept
discipline with a horsewhip, and if his behaviour towards his
grandchildren is any measure of how he treated his own then he
was clearly a great tease. One of his grandchildren, Alf
Hudspeth, told me his favourite tease was to agree with everything the children said. 'Did you work on a treadmill, Granda?'
'Yes, my boy.' 'Did you live with gipsies, Granda?' 'Yes, my
boy.'

Norfolk John cultivated a public image of himself as something
of a character. His Sunday best was a tailed coat and bowler
hat. He joined the village life but in a sense, too, kept himself
apart. He acquired some notoriety for building a cart for his

horse in the upstairs bedroom of the cottage and on completion
not being able to get it down the stairs. My grandfather used
to insist that this often-told tale was not true but that his father
let it go unchallenged for a bit of fun.

His family often experienced his humour to their discomfort.
My grandfather used to tell the tale of how, recovering from an
accident in the pit, his father had to use crutches for walking.
The family felt that, after a few weeks, the old man had no
further use for the crutches yet he still would not part with
them. They thought, as my grandfather put it, he was 'shammin'
to avoid going back to work or doing any work around the
house. One day, however, looking over a sick pig, he poked
the animal with his crutch and the pig turned and bit him on the
leg. At this he began to flay the animal with his crutch, chasing
the poor creature around the field as it fled, squealing, to
avoid his blows.

Even as a very old man he retained his eccentricity. My uncle
Bill likes to relate how, on a visit to Mount Pleasant, Norfolk
John was told that there was a goat in the allotment that was not
wanted. Not wanting to miss a chance, he took the goat on a
rope from Throckley to Eltringham on the other side of the Tyne.
He was then 82 years of age and without thought of his age
or his stomach he walked away, the goat dragging behind.

To live frugally may have been stressed as a virtue in Victorian England and it was certainly a sentiment echoed loudly
from the pulpit of Heddon church, but it was also something
which my great-grandparents could not avoid. Yet the family
produced a lot of its own food. There were chickens and pigs;
this meant a home-based supply of eggs and bacon. The cow
supplied milk and the garden gave them vegetables. There was,
in addition, the payment in kind they got from the farm for
casual work - potatoes, turnips, milk. Finally, there was the
food which grew around them. My grandfather used to say that
nobody but the Browns ever got a blackberry or a mushroom
from Heddon Common. The whole family scoured the place
regularly; what was there they got, and often they got enough
both to meet their own needs and to sell. They were self-
sufficient in jam.

Their income was augmented by the horse and trap and the
part-time carrying business which the old man ran, taking
people and post between Heddon and Wylam and sometimes going
as far as Newcastle. During the summer the Browns took their
horse to Ryton Willows, a pleasure spot on the south side of the
Tyne, to give pony rides at tuppence a time. They bought and
sold livestock; the children, even when they should have been
at school, often worked on the harvest.

In the time they had for play the common, the valley and the
river below offered rich opportunities. This was the same valley
etched for posterity in the woodcuts of Thomas Bewicke. When
my grandfather was a boy the river was rich in salmon and
trout; they fished it, swam in it, lit fires by the side of it and

walked on roughly cut stilts across it. By its banks they went
birds-nesting, stealing the eggs of sand martens, moorhens
and the occasional kingfisher. In the woods around Close House
they found tawny owls and magpies. There were badgers, too,
and deer and rabbits in abundance.

Their sense of their own locality was shaped by how far they
could reasonably walk. They walked to Throckley and Newburn
and back up the tidal stretch of the Tyne to Wylam, the birthplace of George Stephenson, the railway engineer whose cottage
still stands by the riverside. They crossed the river by rowing-
boat ferry to Ryton to play on the open land of the 'Willows'.
They scrambled up pit heaps and watched engines on the dilly
line pulling endless tubs of coal. Visiting fairs, seasonal events
like the picnic at Hedwyn Streams, the hunt, farm sales, concerts, cricket matches and football games offered diversion from
the compelling commitments of work in the house, school and,
of course, a little later, the pit itself. They particularly liked
a pig killing. Several families in the village kept pigs and the
children had great fun evading the adults to get illicitly close to
the killing of a pig; here was rich excitement.

The boys frequently got into trouble although never seriously.
The risks of trouble came directly from their play. Water bailiffs
policed the river, game-keepers stalked the woods. School
attendance officers visited families and the colliery police kept
a sure eye open for trespass and petty theft. And because they
were, like other children, well known, they knew that anything
they did was likely to be reported. But the Browns were not
thought of as troublesome. They were regarded as odd on
account of the size of their family, and Norfolk John was thought
something of an eccentric, but the family was accepted; they
were 'rough diamonds' whose encounters with authority were no
more serious than anyone else's.

Their position was a lowly one but they had their pride. Their
children were carefully instructed that they were no better and
no worse than anybody else. They did, of course, feel their low
status; they were not invited to the garden parties at the
vicarage and the contrast between their home and those of the
gentry was a stark one. Their furniture, for instance, was
simple. And as if to emphasise their poverty, but actually saying
a lot more about the standards the family were subsequently to
acquire, my aunt Eva told me that 'they never had tablecloths,
you know. They ate from a scrubbed wood table'.

They held cleanliness as a special virtue. Eva stressed for me
that her grandmother was ashamed of nothing: 'Even if King
Pete came she wouldn't put off her washing - dozens of white
pinnies hanging on the line.' The impression left with me,
having discussed my great-grandparents extensively with
relatives, is of a couple burdened (though not bowed) with
children, living off their wits, totally untouched by deference
and living very much for the present. They themselves had
no strong attachment to Heddon although their children,

growing up there, saw Heddon as their home and where they came from. They were largely uncritical of the world around them and utterly unconcerned with politics. Norfolk John was a quick-tempered eccentric and congestion in the home often led to angered outbursts. It was, none the less, a close family; if their feelings erupted quickly they carried few grudges. They were all too busily concerned with the present to bother. They knew their place, but since they expected no other that did not worry them either.

2 SCHOOL AND INTO THE PIT

Norfolk John and his wife were barely literate and books played
no role in their lives. They did, however, like most other Heddon parents, make an effort to send their children to school.
It was, in any case, a convenient way of getting the older ones
out of the house, an important consideration in an uncontrol-
lably growing family.

Heddon school was well supported by local employers. It had
been built in 1851, paid for by Mrs Bewicke, although its earlier
history can be traced back to the 1820s and there was a long
if precarious tradition of school attendance in the village. From
1870 onwards the school was under the jurisdiction of the Newcastle School Board and, indicating its strong links with the
established church, the rural dean of Corbridge. Without the
financial support of its local benefactors the school would have
been insolvent. School pence and Board of Education grants
accounted for only two-thirds of the expenditure of the school
(Williamson, 1980).

My grandfather's education was a major factor shaping his
class position. As I have shown elsewhere (Williamson, 1980),
it imbodied massively dominant assumptions about the status of
working-class people, conceiving no other role for them than
that of subordinate workers. My grandfather took to his grave
the idea that he was no scholar. He got little from his education
and expected even less, and throughout a long life he never
really changed his view that being able to work hard mattered
more than 'book learning'. At the same time, however, he never
thought of himself as a 'dud' or a failure. For his generation of
pit lads, doing well at school or doing badly were not things
that mattered much. Joe Robson, a Tyneside song writer, sums
this up nicely in his song, The Pitman's Happy Times:

> We didn't heed much lairnin' then,
> We had ne time for skyul;
> Pit laddies work'd for spendin's sake
> An' nyen was thowt a fyul.

With the migration of farm labour, the availability of casual
work on local farms and the prohibitive cost of school attendance,
especially during periods of industrial strife, it is hardly surprising that Heddon school, like many others in the north-east
of England, had many problems with attendance. J.R. Blakiston,
the chief inspector for the area, complained in his annual report

in 1886: 'There is strong reason to believe that there are still
many thousands of children over the age of six, and even older,
who have never been inside a schoolroom' (Blakiston, 1886,
p. 263). His colleague, Mr J.L. Hammond, in evidence to the
Schools Inquiry Commission just a few years earlier, perhaps
gets to the root of the problem, when, referring to schools for
working people in Northumberland, he said:

> Practically in these schools nothing is taught beyond read-
> ing, spelling, cyphering and writing.... Intellectually con-
> sidered, the instruction given at these schools is extremely
> meagre. In fact, no mental faculty of the pupils is exercised
> or even interfered with by the teacher. (Hammond, 1867-8,
> p. 276)

There were other reasons which account for the low priority the
children of Heddon accorded schooling. The curriculum was
severely narrow and saturated with the values of patriotism,
of Empire and, through the poetry of Goldsmith and Mrs Hemans
- selected by the Board of Education itself - a mild rural nos-
talgia.

The children were taught by poorly trained and nervous
pupil teachers - the front line troops of Victorian education -
within a stern and disciplined order presided over by the head
teacher. Punishment at school was frequent and the Brown boys
were regularly in receipt of it. My grandfather said he was
caned many a time. There are no surviving records of this in
the school log books which remain but his brother Harry's
exploits are regularly listed: 'I punished Harry Brown and John
Laws for climbing the playground walls' (Heddon-on-the-Wall
School Log Books, 30 October 1888); 'Punished Harry Brown
for insubordination' (ibid., 10 June 1890). Harry was eventually
expelled for a year and had to find alternative schooling in
Horsely, a few miles up the valley. The log reads: 'I punished
Harry Brown for stone throwing. He retaliated by wounding me
in the leg, and on being brought before the magistrates his
father was bound over to answer for his good conduct' (ibid.,
12 February 1894).

My grandfather used to say that if he did not feel like school
he just stayed away 'playin' the wag' and keeping well away
from 'the school board man'. It is clear from the log books that
given the choice of a run with the Heddon hunt, a farm sale,
a pig killing or a funeral, certainly casual work of any kind,
and school, the children would 'play the wag'. The abolition of
school pence in 1891 helped matters a little but the chronic
problem of erratic attendance remained.

Pierre Bourdieu (1974) has noted that working-class children
'interiorize their fate' in the course of their education and come
to see education as having little relevance to the likes of them.
Reading the annual reports of the school inspector it is tempting
to suggest that the absolute irrelevance of what the school

taught for the work the children would do on the farms, in the
pit and in the home, together with the fact that they did rather
badly at it anyway, persuaded them that school offered them
very little indeed. And in school itself the children met with
adults who were quite likely to disapprove of them. As to their
performance, the inspector noted in 1887: 'The Elementary
subjects were decidedly below fair.... English was a failure,
as the Repetition was neither accurate nor intelligent and the
Grammar generally weak' (Heddon-on-the-Wall School Log Books,
1887). An earlier report to the Education Commission of 1861
by A.F. Foster concerning education in County Durham perhaps
gives a clue to the way in which public educators viewed the
working classes. Writing about Sunday Schools, Foster notes:
'the teachers conducted their earnest catechising, and the
pupils their eager and intelligent answering in one of the most
uncouth dialects it was ever my lot to hear' (quoted by Seeley,
1973, p. 328). Like many subsequent educators they equated
dialect and forms of speech with intelligence and potential so
that deviations from educated styles of expression were taken as
evidence of failure. As they were not expected to achieve much at
school, it is no wonder that school made little claim on their
time and their interests.

It is one of the nicer ironies of Victorian education that its
pervasive assumptions about order and control manifest themselves in poorly equipped schools with untrained teachers who,
judged by the Heddon experience, had little effect on the moral
or intellectual development of their charges.

My grandfather's real education did not take place in school.
His time there was too short and his experience of it was totally
unconnected with the future held out to him by the pit. Much
later in life he would acknowledge the importance of education,
but in the rare moments when he reflected on his own schooling
he recalled most readily the times when he played the wag. The
only ascertainable residue of his schooling was a thorough
knowledge of popular hymns, something, as will be seen, he
retained to the day he died.

INTO THE PIT

My grandfather left school in 1883 to start work at Heddon pit.
He was 11 years old. He could read and write; he knew a little
poetry, some geography and he had passed the test which
allowed him to leave school and start work.

A raw-boned, slightly impulsive boy, already quite tall, he
was eager to get his 'start' at the pit and earn some money. He
was keen to please, willing to work and prepared to do as he
was told. His father, in any case, was there to see that he did
just that.

To understand what starting at the pit meant to him it is
necessary to stand back a little from Heddon and, indeed, from

the Brown family and see in his starting work the unfolding of
a social process of labour recruitment and work training. I shall
call this process 'pit hardening' to underline the fact that
becoming a pitman was not so much a matter of acquiring particular technical skills – although clearly that was involved – but
of assimilating certain special attitudes and dispositions towards
work which mould the character of miners, setting them apart
from others and without which it would be impossible to work
underground. These attitudes include a strong attachment to
the idea of being 'tough', of not worrying about danger; they
are extremely fatalistic attitudes which allow men to believe that
they themselves are not really at risk, yet if they are to have
an accident there is little they can do about it anyway. Then
there is the value, central to their masculine self-image, of hard
graft and a belief that only 'real men' are capable of it. Like
many children my grandfather must have anticipated such basic
attitudes while at school, but there is much about the mine
which cannot properly be imagined from without, and what it
might really be like underground is not something boys could
properly anticipate.

Pit hardening must be seen, like education, as a process
further defining the miner's class position. The structure of
ownership of the industry, the wage contract and the authority
relationships of the pit itself defined the objective conditions of
that class position. But education and pit hardening defined
how it was experienced subjectively. Through both processes
miners came to see themselves and feel themselves to be miners
and to accept – although never completely and not in an unqua-
lified way since their trade unionism testifies otherwise – the
general social expectations attached to their status as mere
workmen.

My grandfather's parents, after their own experience in rural
Norfolk, urged him to the pit. Farm work for them had a totally
negative connotation. Looked at as a straight economic choice,
the pit was a more attractive proposition than farm work. Wages
were higher and the prospect of regular employment in the
same place greater. In any case, he could start work sooner in
the pit than he could on the farms. The work itself was worse
but the hours were shorter and since he, like thousands of
others, was not encouraged to have any high expectations about
the quality or kind of work he should do, the nature of the work
itself could hardly have been a decisive factor.

The pit was at the foot of Station Bank, tucked into the bottom
of the valley and surrounded by trees. To get to it the pitmen
had to walk a good mile, following a gently curved track. From
the valley top they could see the river cutting its way into
industrial Tyneside. Opposite lay the pit at Clara Vale. To the
left they could see the chimney and winding gear of Isabella pit,
its coke ovens and wagon ways, and the terraced rows of
houses which the coal company had built for its men. By the
river they could see the coal staithes at Ryton Willows and if

the mist had cleared they could make out pits at Lemington, Ryton, Stella and Prudhoe, mounds of waste and red-brick buildings, smoking, noisy and brutally inconsistent with the valley they had scarred and undermined. The Margaret pit was connected to the village by a tramway for tubs which hauled coal up the hillside to a depot where it was stored. Beside the pit were the brickworks, also owned by Mr Bates.

To get to the pit my great-grandfather used his horse; if his shift was different from that of my grandfather they used to arrange that the old man would have the horse for his journey back up the hill; my grandfather would ride it there and his father would ride it back. Like all boys, his first job was to sit by a gate underground, opening and closing it as ponies and tubs went by. Such gates regulated the air supply and the boy would sit there for his ten-hour shift in almost total darkness. The job was a vital one; the careful regulation of air flows prevents dangerous gases building up and so reduces the risk of explosions.

By the time he was 13 he was a driver working with the pit ponies. He enjoyed working with horses and felt confident in his job although it could be dangerous. One of his school friends was killed by a pit pony, and some of them could be very flighty; but he was good with horses, having been brought up with them.

Feelings and expectations are closely interwoven and there is no record of his early start in pit work creating in him any feelings of dread or despair. In fact, such evidence as there is suggests he was keen and utterly unperturbed, despite the fact that he faced a ten-hour shift underground. Indeed, it is likely that he felt quite elated at the idea, for starting work was the essential first step to becoming a man and would confer on him a new position of authority in the family and give him a lot of freedom from home. Some evidence of this feeling of excitement comes from Jack Lawson's biography (1932) and from George Parkinson's account of his childhood (1912). They can be cited here as first-hand accounts of starting at the pit which symbolise the experience of thousands of other boys although, clearly, it would be quite impossible for young boys to articulate their feeling in quite the same way. Nevertheless, such accounts are valuable in that they sharpen the images through which we imaginatively reconstruct the past and they enrich our ability to grasp sympathetically the daily experiences of ordinary people.

Jack Lawson (Lord Lawson of Beamish) recalls his first day at the pit vividly. At five in the morning, 'wedged in between two brothers', he was woken up from a half sleep by the caller: 'Up, up. Get up, Lad. Away lad. Aw-a-a-ay.' And he explains:

The sensation of the traveller who starts on his journey to Central Africa is nothing compared to the thrilling realisation that I was commencing work in the mine that day.... I wanted to see that Aladdin's cave, the pit. (Lawson, 1932, p.45)

The walk to the pit with 'the ring of heavy shoes' around him
seemed to take an eternity. The steel superstructure of the pit
head overawed him a bit, especially since, being small, he was
having great difficulty in keeping his lamp from trailing on the
ground:

> There was steel everywhere. We were surrounded by it; we
> could hear it in the crashing coal 'tipper' and running-tub.
> We saw the thick, glistening, steel-like ropes gliding up and
> down the shaft and the steel chains emerging, heralding the
> coming of the steel cage which carried the iron shaft gates
> upward in its flight. (ibid., p. 45)

He says he 'shrank inwardly' at this 'but the lure of the pit did
not diminish in the least'.

> Its mystery called and drew me like a magnet, and I was
> thrilled when at last I found myself with some forty others,
> sliding slowly and silently down the deep shaft. The slimy
> beams at the side, the black depths I could glimpse, and the
> flashing lights of a seam we passed, all held me spellbound.
> (ibid., p. 46)

But it was not just the wondrous experience of the pit which
enthralled him. It was the change of status his starting work
implied which was by far the most significant feature of it. 'Now
that I was a wage-earner I could go out at night for as long as
I liked and where I liked. Thus ten hours a day in the dark
prison below really meant freedom for me' (ibid., p. 47). This
odd paradox is also recorded in George Parkinson's account of
his own descent into the pit although he writes of an earlier
period, the 1830s. He, too, needed no rousing out of bed. He,
too, welcomed the transition to manhood:

> I looked down with pity on the poor boys who had to continue
> at school and struggle on with vulgar fractions, whilst I
> should not only earn some money but be initiated into what
> seemed to me the mysteries and the manly phraseology of a
> pit-boy's life. (Parkinson, 1912, p. 16)

His mother's advice to him as he left the house (and for the
mother to get up with the men was the common practice arising,
at least Jack Lawson suggests, from the primitive fear that this
parting might, indeed, be the last one) was simple: 'Be very
careful, hinney and mind what thi father says.' He, too, was
impressed in the morning half-light with the 'grimy massive
woodwork around the pit's mouth ... the clanking of engines,
the creaking of the pulleys overhead, and the running of the
ropes in the shaft'. He was terrified by the 'terrible depths of
darkness' into which he was about to descend and the only
comfort he got from his father was, 'Keep thi heart up, hinney;

thoo'll mak' a good pitman yet.'

The work of the trapper boy spanned two shifts and as his father left the pit that day, leaving his son at the gate, he said with tears in his eyes, 'Aw wish ye'd byeth been lasses.' And Parkinson says, 'a feeling of loneliness came over me.... As I looked on the wall of coal before and behind me, and on the roof overhead, home and friends seemed a long way off in the world above' (ibid., p. 23). Both men were struck by the unique aesthetic of the pit, the glistening steel, the noise, the power, the contrast of light and day, darkness and candle-light. Parkinson says of his trapper boy's candle, 'To this moment I have never seen candles burn so brightly.' And of the quietness and solitude - something which to the surface worker might seem terrifying - he has this to say:

> During a long silent interval my candle went out, and, alone in the darkness which might almost be felt, I sat in my hole afraid to breathe. The fearful silence grew oppressive, till I noticed for the first time the sounds made by the gentle ooz-ing of gas and water escaping from the close grained coal around me. A strange and harmonious combination of soft and pleasant sounds they made, delicately varied in tone, rising and falling, now feeble and now full, occasionally ceasing as if their force were spent, then again chiming in perfect con-cord. All the sounds, though independent of each other, combined to form a symphony which seemed very beautiful to the lonely trapper boy. (ibid., p. 23)

Similar feelings are described in the coalfield novel by Harold Heslop, 'The Earth Beneath' (1946), which was very popular in south-west Durham in the 1940s, its popularity deriving clearly from the way miners could recognise themselves and their history in the story of the Akers family which is the core of the book. Of the experience of George Akers going down the pit for the first time Heslop, himself a pitman, writes:

> He discovered a thousand silences in the mine. There, all about him, were the dense, loud silences that he had to learn to recognise.... Each silence had its genesis some-where in the loud dark roar, and each silence was different from the other. There was the silence of the roof. That was the most menacing. It hung there in great slabs of grey-blue shale, a continuous and awful silence, ever-muttering in far-away corners, always ready to leap out of its grim tidiness to become a tearing menace seeking to devour and to destroy. (Heslop, 1946, p. 27)

And of the darkness he says this: 'There is no infinitude like the darkness of a mine, nothing so obscene, it oppresses every nerve in the body. It is the absolute' (ibid., p. 25). Akers's parents may have viewed his going down the pit more philo-

sophically. Work in the pit was dangerous. They themselves
had not been bred to it. It was well known, and had been so
since 1842, that young boys in the pit not only ran considerable
risks of accident but could often cause accidents. Fatal accidents
were a regular occurrence but the risk of non-fatal accidents
was also high. Cuts and bruises were common enough. The
movement of tubs and ponies on narrow tracks always carried
risks of broken limbs, jammed fingers, crushed toes. Roof falls,
explosions and flooding could all occur away from the stalls in
which the coals were actually being cut. A very moving letter
to the local newspaper by a miner from Throckley emphasises
that the necessity of having children working in the pits stirred
not only pity and suffering but also anger in their parents.
Complaining about the way in which a recent correspondent,
referred to as 'Nondum', had misrepresented the miners in the
paper, the pitman from Throckley urged all those who criticise
miners to go down the pits themselves for they would then see
that this kind of work was not something human beings could
just accept. And in a specific reference to child labour he
wrote:

Now, Mr Editor, I will ask 'Nondum' whether it would but
wring even his heart with agony to see an offspring of his
own lifted (more than half asleep) out of bed at three or
four o'clock in the morning and be doomed for fourteen long
uneasy hours to toil with aching limbs and dwarf his young
mind in the recesses of a coal mine, never seeing the light of
the sun between weekend and weekend; would 'Nondum' but
fancy that even Hope itself would be stifled in the labour of
a boy at such a tender age? Would it not wake even the cold
heart of the most feelingless to cry for justice...? We need
not thank 'Nondum' and his class that our boys are now
protected by law from rusting the budding intellect in inky
darkness for so many hours at a stretch.... ('Hexham
Courant', 23 August 1873)

There is a big difference, however, between reflecting in the
abstract about the dangers and injustice of children in the pits
and actually facing daily the prospect that a child might not
return home or, if he did, he might easily be badly hurt. It
simply did not do to dwell on such prospects. All that could be
done was trust to luck and the Lord himself and get on with
what had to be done, careful not to communicate those dark
fears to the child. The busy-ness of daily routines has a
soporific effect; to be occupied is a way of avoiding contemplat-
ing the unthinkable. The thought that there were other children
in the pit must have been a great source of comfort to my
grandfather's parents; the risks, pain and guilt attached to
sending him there each day could then be shared with everybody else.
    The worry of it all, however, is captured movingly by Joe

Skipsey, the pitman poet from Northumberland, in his poem,
Mother Wept.

Mother Wept

Mother wept and father sighed:
With delight aglow
Cried the lad, 'Tomorrow',
'To the pit I go.'

Up and down the place he sped –
Greeted old and young;
Far and wide the tidings spread;
Clapped his hands and sung.

Came his cronies; some to gaze
Wrapped in wonder; some
Free with counsel; some with praise;
Some with envy dumb.

'May he', many a gossip cried,
'Be from peril kept',
Father hid his face and sighed;
Mother turned and wept.

(From *Joe Skipsey, Pitman Poet of Percy Main*
(1832-1903))

My grandfather's early experience at the pit, as it was for
thousands of others, was an induction into the adult world he
himself would inhabit. The adults he saw around him were the
sort of man he himself would come to be, and in watching them
he would gain some appreciation of his own future; in living
through the tragedies he would gain some sinister clues to the
kind of world he lived in. An old miner and lifelong friend and
workmate of my grandfather, Mr Stobart, told me that he would
never forget, as a boy, seeing a Mr MacDonald being carried
out of the pit on a stretcher. Both his legs were broken and
there were fears that his hips and back might be too. As he
was carried out he was muttering over and over again, 'Whaat's
gan 'i become o' me bairns?' As Mr Stobart said (he had just
celebrated his hundredth birthday), 'I nivver forgot that ...
nivver.'

Worries about the effects of accidents on the family income
might have been assuaged a little in the Brown household by
the sheer size of the family with several boys working, but they
could not have been complacent about it. Under the Employer's
Insurance Liability Act of 1880 coal owners were required to
insure their workmen against accidents and loss of life. The
Steam Collieries Defence Association (later the Northumberland
Coal Owners Association) set up their scheme with Thos. Bates

signing the agreements for Heddon colliery. The minutes for
11 March 1881 set out the rules governing the 'charities' of the
association. 'Smart money', it declares, is to be paid to injured
miners 'engaged in any way to the advantage of the owners;
and in all cases, except at Heddon, it is paid for injuries
received while travelling from face and bank'. Money was payable for 'beat hands damage from fellow workmen's picks,
sprains, carelessness'. In the case of fatal accidents, collieries
provided coffins and £1 towards funeral expenses. They also
undertook, as required, to provide a man and a hearse for the
funeral itself. The scheme was financed at the assurance rate
of one shilling per ton of coals drawn from the colliery. Minute
book no. 2 of this association sets out many settlements which
took place under the scheme, indicating gravely, though not
intentionally, the haunting fears from which no mining family
could be entirely free. Thus:

> No. 334 Heddon. William Breckons, 14, driver.
> Killed on July 12th 1906, by a fall of stone.
> Family consisted of the following:

	Age	Earnings Per Week	Remarks
Deceased	14	£0.8.0	Gave all earnings to parents
Father	42	1.2.0	
Mother	34		
Brother	16	0.8.1	
Brother	10		

> Claim made by father for £40
> Awarded £25

This particular entry refers to a boy related to my grandfather.
William Breckons was his nephew, the son of his sister, Alvina.
Or again:

> No. 571 Heddon. Robert Lowney 33, deputy.
>
> Killed on January 13th 1911, by a fall of stone.
> Deceased leaves widow, two daughters aged 10 and 7
> and a son aged 1, wholly dependent.

Three years earnings	£322.4.0
House and Coals	39.4.0

> Full liability admitted £300

> (Northumberland Coal Owners' Mutual Protection Minute
> Books, NRO 60 NCB/DL/L)

The first point about such entries is the meagre provision
they indicated. Families faced with the loss of the principal
earner were faced with desperate problems. But the main point,
for me, is the way they underline the proximity of death in the
pit, and in this lie the roots of both fatalism and hedonism
which have been so much a part of working-class life, particularly in mining districts.

My grandfather learned early on, as a boy undergoing his
pit hardening, that it was futile to worry about accidents;
that while much could be done to avoid them there was none the
less nothing to be done against the caprice of fate. And if the
future - at least in so far as it revealed itself in the lives of
those around him who were older, threatened to be bleak - then
it was much better to think on the pleasures of the present and
to have a good time. Just as it says in the song:

> Let's not think on to-morrow
> Lest we disappointed be
> Our joys may turn to sorrow
> As we may daily see.

Pit hardening was therefore not just a technical business of
finding out how pits worked or what the job of a pitman was.
Indeed, in giving evidence to the Select Committee on Mines in
1866 Thomas Burt, described as a coal hewer but a man later to
become a miners' MP, and at that time a union leader, pointed
out quite emphatically that there was no difference in skill
between a pitman who had worked as a boy in the pits and someone who had come in as an adult ('Report from the Select Committee on Mines, Minutes of Evidence', 1866, 1-15). The employment of children could not be justified on training grounds. It
was more a subtle process of getting boys used to a whole way
of life and to coping with, through suppression, those fears
which, if allowed out, would prevent a man ever going underground. The irony, even the tragedy, of it all was that coal
companies could actually rely on some of the attitudes of the men
themselves - their sense of manliness and their unwillingness to
think too far ahead, their toughness and their acceptance of
work, any kind of work, as their own peculiar fate about which
it did not do to grumble too much - to reinforce the responses
from young boys which the company required of them as
employees. These responses include regular attendance, an
acceptance of authority, sustained effort at work and an almost
total acceptance of the conditions of work itself, a respect for
the dangers of the pit but without too much concern about them
and, finally, a sense of resignation that since this was what pit
work was like it was not worth bothering too much about changing it or dreaming of other things. That, in the end, is the full
tragedy of pit hardening; it closed off their dreams and trapped
them in the present tense.

HEDDON PIT

Heddon pit was known locally as a bit of a 'blackening factory'
- that is, as an old, not very efficient pit. The drawing shaft
was narrow at the top and bottom and wide in the middle. The
men entered it through a drift tunnel which spiralled downwards
into the earth, the last few feet of the drift being on a very
steep incline known as 'Knack Bank'. The winding engine was
a single-cylinder machine which regularly jammed and had to be
rocked to jerk it back into action.

The men who worked in the pit, so far as can be ascertained,
originated in the north of England and particularly areas in
Northumberland close to Heddon. Of the coal miners listed in
the 1871 census enumerator sheets for Heddon (twenty people
in all) only three had been born outside the county. The pit
workers, then, were local people. Its owner, Mr Bates, was well
known in the village. Its managers, living beside the colliery,
were clearly seen as part of the village itself. They were
accorded respect and they were not feared in any way. Indeed,
there were times when the pit manager seemed little different
from the men he employed. Mrs Hall, a lifelong resident of
Heddon (born 1894), recalled how, before the First World War,
Fenwick Charlton, one of the colliery officials, actually broke
into the village chapel: 'Him and Joby got too much to drink here
and they broke into Heddon Chapel. They cut themselves with
glass and stuff and they had to buy a new bible for the Chapel'
(NRO T/114). And of the manager, Mr Musgrave, she said, 'He
joined in with the village.' The pit, then, was not an impersonal
place.

There was, however, a clear hierarchy among its employees.
The managers and officials, as in every other mining village,
had higher wages but, in Heddon, not necessarily better houses.
There were no special houses for officials and, indeed, many of
the miners themselves rented their homes privately from local
landowners. In this respect Heddon could not be described as a
typical mining village; there were too many opportunities for
work outside the pits and for houses to be had outside the control of the coal company. But the managers did get a better
class of coal as part of their wage. Miss Elliot, another lifelong
Heddon resident (born 1892), noted in an interview that her
father being a 'master's man' got the 'second quality coal'. 'He
didn't get the small like the miners' (NRO T/117).

The tone of working relationships in the pit can only be
guessed at. E. Dückershoff, a German miner who worked in the
area in the 1890s, noted a particular spirit of comradeship in
the English pits:

> In the pits here there is a spirit of comradeship. Every order
> is given and carried out in a friendly manner. Cursing and
> bad language are seldom heard; here it is a real pleasure to
> work. (Dückershoff, 1899, p. 13)

The working hours, from 10 a.m. to 4 p.m. and 4 p.m. to
10 p.m., he thought were fair and manageable, and with wages
at 5s. 6d. per day, each payment being made fortnightly,
Dückershoff was impressed by the comparative affluence of the
miner's life:

> He gets up about eight o'clock and breakfasts on bacon or
> brawn, with a couple of eggs, and bread and tea. He takes
> a couple of slices of bread and meat or cheese with him to
> the pit. On finishing the shift at four o'clock, he has meat
> and pudding, or soup with eggs or meat and for supper,
> bread and cheese or meat, with tea, the kinds of meat
> always changing. (ibid., p. 31)

These remarks are particularly interesting since they come from
a foreigner. He is curious about everyday things, and since
these are seen in a comparative perspective they are made to
stand out. The healthy diets at this point reflect good wages
and contrast sharply with the intense exploitation Dückershoff
claims to have experienced in the German coalfields. German
industrialisation was in his view being paid for dearly by
German working men. But he was not so impressed with the
Englishman's politics:

> English workmen do not attack capital itself but only the
> nuisance of capital, and the exploitation of workmen by the
> capitalist class. In point of economics, the English workman
> is in advance of the German. In points of politics, he is
> behind him. (ibid., p. 78)

The German may well have been influenced by marxism. In
Lemington where he was working, as in Heddon, and, indeed,
throughout the coalfields, the predominant political mood was
Liberal.

Until 1918 my grandfather was a Liberal. The MP for the
Wansbeck division (the parliamentary constituency), Charles
Fenwick, was a Liberal. The Northumberland Miners' leader,
Thomas Burt, elected to Parliament in 1874 as a 'Radical Labour'
Member and a man who straddled coalfield politics like a colossus
for over forty years, was a Liberal. Page-Arnot, the miners'
historian, says of Burt that he

> embraced the Liberal creed with an intellectual fervour that
> led him to accept all its current applications, not only to
> politics but to industrial problems. Hence his view on the
> identity of capital and labour was part of that contemporary
> political economy which he both practised and preached.
> (Page-Arnot, 1949, p. 53)

Burt was also a deeply religious man, a Methodist, just like his
colleague John Wilson, secretary of the Durham Miners' Associa-

tion. He was active, too, in the temperance movement. Throughout his political career he had pressed for the reform of Parliament and the improvement of the living conditions of miners. He
did not, however, support the movement from the 1880s onwards
for the eight-hour day legislation; he resolutely held out against
joining the Miners' Federation of Great Britain on the grounds
that national unions should not interfere with the mechanism
of the sliding scale which ought to govern wage negotiations
between men and their employers. Burt preferred instead the
strengthening of local conciliation machinery and arbitration.
This belief of his, which permeated the leadership of the
Northumberland miners, was at the root of what Sydney and
Beatrice Webb thought was the acceptance by Northumberland
pitmen of the social and economic views of the mine owners. The
'victory of arbitration', said the Webbs, 'brought results which
largely neutralised the advantages'. As in the case of political
triumphs, the 'men gained their point at the cost of adopting the
intellectual position of their opponents' (quoted by Page-Arnot,
1949, p. 125).
   Dückershoff was therefore highlighting a vital feature of coalfield politics. Unwittingly he was also saying something important about the differences which separated the labour movement
on the Continent from that in Britain. Burt and Fenwick had
been active in the late 1880s in setting up an international
association of miners. At the first meeting of the international
conference in 1890 (a meeting decided upon at the first Socialist
International in Paris in 1889 which itself had been convened
to celebrate the centenary of the French Revolution) the British
delegation (at Jolimont in Belgium) were shocked to discover
that the trade union organisation among Continental miners was
very weak although their socialist rhetoric was very powerful.
In Britain the opposite was true; organisation was strong and
socialism undeveloped. British miners were, therefore, always
more likely to press for changes in their conditions through
industrial and parliamentary methods; on the Continent, and
particularly in Germany, miners were far more interested in
revolutionary marxism.

THE HEDDON LODGE

There are no surviving records of the union at Heddon, so it is
only possible to gain glimpses of what preoccupied the men in
the pit. There was sufficient solidarity among them socially to
enable them to organise a rota system for calling each other up
in the mornings. The one on duty each morning used to stand in
the village square and yell to the rest to get out of bed, and
there was, as I have already shown, a rich associational life in
the village which reinforced common marks of identity in the pit.
The politics of the miners' union leadership in the period up to
the turn of the century and before the First World War may

have been Liberal but many of the Heddon pitmen could well
have been Tory in their outlook. Mrs Hall, the old lady to whom
I have already referred, speaking about pre First World War
elections, explained:

> Newburn was a Liberal place. Heddon was a Tory place. The
> better class was all Tories and the farmers. My father was a
> pitman but he was a staunch Conservative. There was a lot of
> pitmen voted Tory. He would have a red ribbon in there. We
> had a dog and it would have a red ribbon in as well. And we
> had to have a red ribbon in our hair.

The lodge leaders at Heddon were strongly connected with the
chapel. George Anderson and Harold Jackson were both chapel
men. The lads used to call Harry Jackson 'Holy Harry'. He was,
however, well respected. As an orphan aged 10 he had escaped
from the Newcastle workhouse, subsequently working in several
pits in the area and eventually settling in Heddon. He was an
early convert to Wesleyan Methodism and he remained a devout
Christian till his death in 1929. He was an active supporter of
Charles Fenwick, the Liberal MP, and was for many years
secretary of the Heddon Liberal Association (see Northumberland
Miners' Association, Minutes, William Straker, Monthly Circular,
1929). For my grandfather, therefore, the link between people
like himself - the poor of the village, in contrast to those in the
social circle of the church - and politics generally was most
obviously made by the Liberals with their close connections with
the union.
In the period from his starting work to the turn of the century, when he left Heddon pit to work for the Throckley coal
company, the problems the lodge had to cope with were largely
falling wages under changing sliding scale arrangements and
redundancy. On a broader plane the Northumberland Miners'
Association (NMA) was concerned in this period with parliamentary reform, housing policy, the rise of a labour politics with
the dark promise of class warfare, whether to join the Miners'
Federation of Great Britain (MFGB), and impending legislation
on the eight-hour day. These are the issues around which my
grandfather's early political education, such as it was, crystallised. These are the items which Anderson, Jackson, J. Graham
and J. Wilson - all of them at various times Heddon delegates to
the council of the Northumberland Miners' Association - would
have reported on at lodge meetings and which were discussed
at home after lodge meetings.
Judging by the minutes of the Northumberland Miners' Association of the period, there was little which would encourage
the miners of Heddon to form a coherent class analysis of their
condition. In a letter to Joseph Chamberlain, MP, in February
1884, for example, the union stressed the need for safety legislation and for parliamentary reform, and underlined their
interest in the politics, not of class warfare, but of self-help.

With, as they put it, 'other Liberals of the North', they welcomed Chamberlain to Tyneside and pointed out to him:

> We belong to an industry which employs half a million men
> who follow their daily avocations at the peril of their lives.
> *We are firm believers in self-help* and have done much for
> ourselves, but we are not altogether independent of Parlia-
> ment for protection. (NMA Minutes, NRO 759/68 my
> emphasis);

And in 1888, in an address to the members of the union, Burt
fired a broadside at the idea of a purely labour party in politics.
He insisted the 'working men did not want class representation
but they objected to class exclusion' (ibid., 1888, p. 2). A
labour party, he argued, would be too sectional in its outlook.
'Our aim should be to unite men, not to divide them; to break
down, and not to intensify and accentuate class distinctions'
(ibid., p. 3). Self-help and consensus politics are hardly the
ingredients of class consciousness.

Nevertheless, the consciousness of exploitation and hardship
was firmly rooted in the experience of working at the pits. Daily
bargaining over prices for different types of underground work
encourages, as I shall show in more detail in a subsequent
chapter, a defensive belligerency which is essential to protect-
ing the level of earnings underground. That such bargaining
could escalate into full-scale trials of strength between the
owners and their men was something my grandfather grew up
with as a fact of life. As a very small boy in 1877 he lived
through the so-called 'Nine Weeks Strike'. In the first year of
his employment at the pit, 1883, he was himself involved in a
stoppage lasting twenty-one weeks when the miners of both
Northumberland and Durham resisted a plan on the owners' part
to introduce a contract system of monthly notices.

It is not inevitable, however, that the experience of such
conflict should lead to a class-conscious outlook. The experience
of the famous seventeen-week strike in Northumberland in 1887
illustrates this nicely. The men were defeated – the owners had
asked for a 15 per cent reduction on the sliding scale and the
men resisted it, having to agree finally to a 12½ per cent reduction – but there appears to have been little evidence of this
protracted dispute creating either bitterness or disturbance.
An editorial in the 'Newcastle Weekly Chronicle' of 5 March 1887
noted, for instance:

> The strike in the Northumberland Coal Trade is pursuing so
> placid a course that few people in the district are able to
> realise that thousands of men have ceased work. Even the
> newspapers have little to say on the subject, for the simple
> reason that there are no exciting incidents to record.

There was, of course, great hardship, especially for non-union miners. A letter, published in the 'Weekly Chronicle', from the Blyth Relief Committee to the coal owners urging them to settle the dispute gives some idea of how difficult things were:

> It is within our knowledge that in many families there is pinching poverty, in some semi-starvation, and in not a few little ones are crying for the bread that mothers cannot buy, and we cannot provide.... The unemployed and poor we know aid each other, but their ability for neighbourly help is now almost exhausted. ('Weekly Chronicle', 26 March 1887)

By this time there were seven children in the Brown family and Annie was just a baby; there were only two wage earners, my great-grandfather and my grandfather. Some Heddon pitmen toured the area busking to earn some money. A 'Weekly Chronicle' report of 16 April describes three Heddon pitmen playing a tin whistle and a banjo in Hexham market place to such effect that they destroyed the trade of the permanent Hexham buskers. The Browns, however, relied on their own resources, their home-cured bacon, their chickens, occasional casual work on Law's farm for which they were paid in kind, and the proceeds from their carrying business.

There was clearly a strong resolve behind this strike and the strike itself was strongly supported by other trade unionists and the public at large. The Newcastle Relief Committee, for instance, collected £2,005 17s. 4d. in aid of strikers and their families. And in the Heddon area a poet pitman and activist, Frank McKay, employed then, although not afterwards, by the Throckley coal company, circulated a poem about that resolve:

> Now, when we've brought our gear to bank
>   and boldly faced the foe
> There must be 'no reduction'
>   or to work we will not go
> We must not let those Bishops
>   or Ministers of the Crown
> Step in to settle our dispute
>   and bring our wages down.
> We can manage all our own affairs,
>   and that we mean to do
> With hearts and hands united -
>   like soldiers brave and true
> We'll charge the enemy right and left
>   and do the best we can
> To retain the present wages
>   for 'The Honest Working Man.'
> ('Newcastle Weekly Chronicle', 15 February 1887)

The wages at issue were 5s. 2d. for a seven-hour shift underground, a rate of just under 9d. an hour. The strike ended in

May with the miners accepting a 12½ per cent reduction in these wages. The leader of the Northumberland Miners, Thomas Burt, had urged his men to settle on the ground that they could never win a strike in a declining market. This same argument was clearly proclaimed by the 'Weekly Chronicle' in an editorial warning the miners of the future ahead of them:

> In the light of recent events, ... it will be well to bear in mind the fact that the age which built up our own trade has also made an open market of the whole country, and that it is simply butting the head against a rock, or an illustration of Quixotic tilting, to fight against the necessary regulation of wages according to ruling values. ('Newcastle Weekly Chronicle' 28 May 1887)

This argument is a vital one for it clearly tries to link the fate of miners' wages to factors over which the miners and, indeed, the coal owners themselves could have no control over whatsoever: the market. In the case of Northumberland coal exported to the Baltic, the market had an additional uncontrollable quirk. Since the Baltic froze over in winter coal supplies and, in consequence, employment in the pits were cut. As the miners themselves subscribed to this argument they would not equate their own exploitation with their immediate employers. The influential 'Weekly Chronicle' reinforced this view and did so throughout the strike. In January, for instance, the miners were told:

> if the miners come out on strike, according to all past experience, they will, after weeks of privation and weary waiting, be in no better position than they are at present.
>   The men say they can hardly live at the present rate of wages, and the owners say they cannot make their collieries pay.

What, then, was the problem? The paper was quite clear that the issues raised in the dispute should be translated into an 'attack on the system of royalties and wayleaves, the owners of which are reaping the harvest denied to workers and capitalists alike' (22 January 1887). The significance of this is that the issue of exploitation is further removed from the employment relationship itself and transferred, if anywhere at all, to an anonymous class of landed capitalists who could not be brought to heel through normal union activity. And seen as a broader moment of historical change, it symbolises that harmony of interests between capitalists and workers which the ascendant industrial bourgeoisie of the nineteenth century used as a weapon against the older landed aristocracy. Along with ideas like self-help, laissez-faire and parliamentary reform, the theory of the harmony of interests between capitalist and workmen had been, at least since the agitation to repeal the Corn Laws in the

1840s, an essential ingredient of what Bendix (1956) has called the 'entrepreneurial ideology' and was used to define the interests and legitimate the position of the industrial bourgeoisie.

Such considerations seem far removed from Heddon, but in fact they are not. I have no record of what my grandfather felt about the 1887 strike, and it may be presumptuous to believe that a 15-year-old boy thought much about it anyway. But his income was at stake; his friends were on strike; and if the men did try to explain it to themselves then they would have taken into account the role which markets, landlords, capitalists and unionists all played in the winning of coal, an ideological matrix which *at that point in time* would not support very radical solutions to the miners' plight.

The experience of a protracted and in the end pointless dispute was, I believe, a formative one. It emphasised his powerlessness and it strengthened a resolve to build a defensive line around his home to protect his family from the vagaries of wage labour. It cultivated in him a feeling that too great a dependence on one source of income was risky. Encouraged by his father, like many young men he learned quickly to appreciate the importance of the garden, the pigs and chickens, and the need to grab at any casual work when the opportunity arose. Without such protection the alternatives in Heddon in difficult times were to receive charity, to work on the land or to migrate.

The two years following the 1887 strike were difficult ones. In 1887 the miners terminated the sliding scale and resorted to direct bargaining with the employers. The whole coalfield was in a parlous state. William Straker, Burt's successor, wrote: 'When the pits reopened after the strike the colliery owners had practically to buy back their trade by selling at ruinous prices, so that it was very questionable whether they did not lose much more than they gained by the lock out' (Straker, 1916, p. 54). Prices did recover until 1894 when they again suffered a decline. In 1894 a conciliation board was set up to regulate wages, but it was abandoned in 1896 due to falling prices.

The effects of these difficulties in Heddon can be seen in the figures for union membership. In 1893 thirty-four men were dismissed from the colliery. In 1894 a further ten men were made redundant through 'bad trade'. In July 1893 the Heddon lodge voted against the Northumberland Miners joining the Miners' Federation of Great Britain and in the same year, with their tail between their legs, they voted against strike action in support of a wages demand of 16 per cent. The membership figures, calculated from the annual balance sheets of the Northumberland Miners' Association, are shown in Table 2.1. These figures show a pit in difficulties, shedding labour at a drastic rate, with only a modest growth towards the end of the century and in this respect being far out of line with the coalfield as a whole. It was during this late period that the Throckley coal company tried to take over the Heddon Margaret.

Table 2.1 Union members, Heddon colliery, 1884-1900

Year	No. of members
1884	26
1885	21
1886	61
1887	–
1888	61
1889	51
1890	96
1891	114
1892	109
1893	105
1894	72
1895	34
1896	49
1897	46
1898	65
1900	61

Source: Annual Balance Sheets, NMA, NRO 759/68.

This period includes my grandfather's growth to maturity in his early 20s, the period which he might well have thought of as the peak of his career when, as a powerful young man capable of a great deal of work, he became a coal hewer. During this time it was not unknown for him to work two consecutive shifts in the pit, covering for his father who for whatever reason was unable to put in his shift. As an old man he often spoke of those times when, learning that his father had not 'turned in', he simply had to brace himself for another seven hours under-ground. On occasion it happened that he had come all the way up Heddon bank and had to take the horse back to the pit to do his father's shift.

By the end of the century the negotiations, which had been going on since the early 1880s between the Throckley coal company and Mr Bates over the sale of the Heddon Margaret, were virtually complete. My grandfather was courting. His home was very congested and relationships between himself, some of his brothers and certainly his father were becoming abrasive.

A persistent theme of union politics at this time was housing. The coal owners had tried on several occasions since 1888, when there had been a strike at Delaval pit in Benwell over rents for colliery houses, to charge miners a rent for their houses. This the union had resisted. The union's aim was to improve the amount and quality of housing for miners, and in 1900 it had passed a firm resolution urging collieries to build more houses and to improve existing ones (NMA, Minutes, June 1900, NRO 759/68).

The relevance of this to my grandfather was that he needed a house. Unlike other villages, Heddon had many miners who had to rent privately, and housing was short. The insecurity of his job at Heddon and the need for better housing eventually persuaded him to move on, to Throckley. From trapper boy, through a spell as a coal putter, he was now a fully qualified hewer. Seventeen years in Heddon pit had helped to fashion his basic outlook and establish his secure self-respect as a good worker. They had instilled in him a work discipline which his father had never acquired and which he himself, as will be seen, never lost. Quite apart from the pit skills he had acquired, he had learned to negotiate his way through the underground price system which paid different amounts for different classes of work; he had acquired, through union membership, a clear conviction that miners needed to be organised to protect themselves. The political stance of the union reinforced his belief that the Liberal Party was for the working man although politics was something only the 'big nobs' (as he used to say) bothered with. Above all, precarious employment, industrial defeats and fluctuating wages had strengthened his determination, so far as it was possible, to be self-reliant and free of a dependence on wages alone.

3 IMAGES OF YOUTH

Outside the pit his obligations to his family pressed hard on
my grandfather, and as he grew older he was expected to do
more - to carry more water, to spend more time in the garden
and with the animals. Seasonal work on the farm - picking
potatoes, grading them, stacking hay ricks - and the regular
work of the farm - milking, grooming horses and cleaning out
the pigs - took up his time although he never found such work
onerous. His friends were from the farm and they all enjoyed
the work.

As he grew the boundaries of his life were extended. On
Saturdays he went with his father to Newcastle with the horse
and trap to pick up orders for people. And when it was clear
that he had full control of the horse he was allowed to take it
away himself, to Wylam to collect the mail or to Newburn to pick
up a parcel. He also took it to Ryton Willows whenever there
was a fair or a meeting there. As it was a popular beauty spot,
visited frequently by tripper parties brought by boat from
Newcastle, there was often the chance to earn a few coppers
giving children a ride on the horse. The Willows (pronounced
'Williz' locally) was an important place for political meetings in
the district before the turn of the century, and in 1887 even
William Morris, the socialist and revolutionary of the Socialist
League, addressed a meeting there.

The long strike of that year brought Morris north along with
Tom Mann of the Social Democratic Federation. This was the
period when the socialists were attempting to build up support
among the urban working class and a time, as I have already
shown, when miners in Northumberland were still strongly
attached to the Liberal Party. Morris described the place,
accurately to anyone who knows it, as 'a piece of rough heathy
ground ... under the bank by which the railway runs: It is a
pretty place and the evening was lovely' (quoted by Thompson,
1977, p. 445). He went on:

Being Easter Monday there were lots of folks there with
swings and cricket and dancing and the like.... I thought
it a queer place for a serious Socialist meeting, but we had
a crowd about us in no time and I spoke, rather too long I
fancy, till the stars came out and it grew dusk and the people
stood and listened still, and when we were done they gave
three cheers for the Socialists, and all was mighty friendly
and pleasant. (ibid.)

Had he cast a glance at the path along the river bank he would
almost certainly have seen my grandfather and his brother Tom
leading children on their horse, earning a few vital coppers
and oblivious to his rhetoric.

There was nothing, of course, in my grandfather's experience
which led him to see politics as having much to do with the
likes of him. And although the union connected questions about
pits with parliamentary politics on such questions as workmen's
compensation, safety in mines and so on, politics was nevertheless something remote and little in his experience could have
given him much confidence that his views mattered. Indeed, they
did not; even the idea that working men had a right to vote was,
in parliamentary terms, a novel one. Parliamentary action by
miners had the character of pressure group politics rather than
that of a movement of working men (see Gregory, 1968).

Information on political questions came to the village through
the 'Hexham Courant' and serious political analysis vied with
the astonished reporting of crime and patriotic accounts of
colonial exploits. Dr Livingstone's dispatches from the dark
continent were extensively reported. Seeing itself as a radical
paper under the sponsorship of Joseph Cowen, it naturally
enough stressed the themes of parliamentary reform and dealt
extensively with local questions and grievances, always reporting at length the public utterances of local politicians. In the
1880s its pages were dominated by home rule for Ireland, parliamentary reform and the issue of religious liberty. Its radicalism
did not, however, extend to a questioning of the rights of
private property; its case rested on the hope that reasonable
men, persuaded to the truth, could be relied on to improve the
lot of others.

Being uninterested in politics is not the same as being unaffected by politics. I have already stressed that during this
period my grandfather's inclinations were Liberal. He was, in
addition, mildly proud of the Empire and sentimentally patriotic
if the occasion demanded it. I suspect, too, that he was mildly
racist. Aunt Eva remembers him once saying that if he was
expected to sleep in a bed that he knew had been slept in by a
black man then he would not do it. This improbable eventuality
creased my mother with laughter; she knows nothing of her
father's attitudes to black people but found it amusing that he
should say such a thing. In Throckley, she told me, 'you never
saw a white face. They were all black from the pit'.

The one political idea which he did retain was that of self-
reliance. The ideological climate of the time allowed a distinction
to be drawn between the deserving and undeserving poor and
between those who were unemployed through misfortune and
those through fecklessness (see Novak, 1978). Self-help and
self-reliance were powerful ideas, suspicious of charity and the
state, and my grandfather accepted them completely. From the
beginning of his working life he paid his weekly insurance into
'Heddon Club', a branch of the Manchester Unity of Oddfellows

Friendly Society. The same spirit of self-protection was what lay behind his membership of the union.

Norfolk John rarely spoke about politics; there is no record of his having instructed his sons in history. What bothered him was the present, not the past. Nothing in my grandfather's schooling reduced his historical ignorance either and, although this must be pure speculation, I doubt whether he had any coherent grasp of the history of the miners. He was not unique in this. In any case, the first history of the miners of Northumberland and Durham, that of Richard Fynes, did not appear until 1873. Heddon itself, being a small village in the relatively undeveloped western part of the coalfield, was insulated from the political currents which elsewhere fuelled a more radical consciousness of injustice.

Death in the pit, news of disasters elsewhere, strikes, though infrequent, could all produce a 'distancing effect' when ordinary men, temporarily lifted out of their routines, could stand back and reflect. Some disasters, like Hartley, were well known. What was lacking, however - or what, perhaps, had been lost - was a coherent political and economic analysis of such events which could lend a radical meaning to experience. And working for a small, paternalistic company in a village where plentiful casual employment on the land took the sharp edge off exploitation, my grandfather was, in contrast to men on the east coalfield or in parts of Durham, not really exposed to a radicalising, historically reflective rhetoric to jerk him to a strong political awareness. Like his father he lived his life rather uncritically in the present tense.

LOCAL LADS

His relations with the village squirearchy were formal and respectful, though not deferential, and largely indifferent. There is only one episode which has filtered down through the family which gives a small clue to this aspect of his early life. Mr Bewicke, the squire, was the first person in the village to acquire a motor car and the fact that it had cost £1,000 had clearly prompted a good deal of local discussion. When the car broke down Mr Bewicke would often call on some of the young men of the village to give him a push. My grandfather's comment about this was always mildly cynical. 'For the want of a few pennyworth of petrol a thousand pound car is no good.' He told this often enough as if to underline the fact that all a man could really trust was his own two legs or a good horse.

What interested him most intensely was his pleasure, and Heddon had much to offer a young man. The Three Tuns pub was the focus of bacchanalian nights boozing, sing-songs and often a battle. Once his 'board' was paid what was left of his money was his own and he spent most of it on drink. If he was not drinking in the Three Tuns then it was likely he would be

at Wylam or even, on summer nights, Horsley on the Hexham
road, although boozing trips away from home were risky; they
often ended in fights with 'local lads' elsewhere (cf. Chaplin,
1978). Such battles, apart from the rich excitement they produced and the mythology they generated about which men were
'the hard men', had the effect of reinforcing an identity with
the village. They were also, of course, a ritual celebration of
the qualities of masculinity which suffused the culture of manual
work.

Doing nothing was a popular pastime; pit lads used to hang
around corner ends, sitting 'on their hunkers' passing the time
just talking. Outside mining communities such behaviour might
have been regarded as pure aimlessness. And they were certainly recognisably different from the less cohesive body of
'hinds' (farm labourers). In Heddon the young lads played
quoits quite a bit behind the pub, but often they just sat and
talked, smoking their pipes, cracking jokes and playing games.
They played 'chucks', a game with pebbles; they played dice,
sometimes cards. I reckon, too, knowing the persistence of the
all-male corner-end culture, they must have enjoyed the wilfully
unsavoury excitement and the bravado of outdoing one another
in telling filthy jokes. The jokes made available to them weekly
by the 'Hexham Courant' seem to me to be far too anodyne to
strike a chord with the pit lads. Here are two of them, said to
be popular in the district in the 1880s:

A country undertaker boasts that he had the best hearse in
the place and defies anybody who ever rode in it to say the
contrary.

A Gentleman passing a woman who was skinning eels, and
observing the torture of the poor animals asked her how she
could have the heart to put the animals in such pain. 'Lord,
sir' she replied, 'they be used to it.'

Closer to their concerns might have been the one which went:
'Wild oats are said to be the only crop that grows by gaslight.'
For it was on the corner end that they learned about sex; here
they could swap tales of their exploits, measure up the girls of
the village and establish their own reputations or perhaps just
learn a bit from the older lads. School ended for most of them
by the age of 12; the corner-end school went on well into adulthood; membership of it was a powerful social marker.

And the group itself, as a source of information, news and
gossip was a powerful instrument of social control among the
lads themselves. The solidarity of the corner end carried right
over into the pit and vice versa. And the group acted as a
reference group; it set the standards of expected behaviour
and demanded loyalty. A pit lad would become so well known by
his mates that he could conceal little from them, even supposing
he wanted to. Pretence was impossible. The group was a critical

forum both supportive and destructive of projected self-images.
The wit, the clown, the hard man and the fool all had their
place. And if it was on the corner end that they traded images
of themselves it was there, too, that they learned their own
history. Images of the past were formed here and passed on.
On the corner end they could talk justice and rights. Their collective experience of the pit and that of their parents could be
filtered and assessed. In the utterly everyday business of
'hevin a bit crack' with their rolling 'r's', their 'thou's', 'thee's'
and 'thine's', a powerful sense of place and position emerged.

The village had a rich associational life, however, which the
lads did join in. There was an annual picnic when the colliery
brass band played and sports were organised for the younger
people. The first annual picnic was in July 1880 and is reported
in the 'Hexham Courant'. In a field 'kindly granted for the
occasion by T. Bates',

> The Heddon Band ... played a choice selection of music dur-
> ing the afternoon, and also at the ball, which was fairly
> attended. Dancing was led off by Mr. Hunter ... to the well-
> known tune of the 'Keel Row' and was carried on until eleven
> o'clock when all quietly dispersed. During the afternoon a
> number of sports were brought off.... (17 July 1880)

As time went on the picnic became much more colourful. Miss
Elliot evoked this when describing the picnic which took place
at the turn of the century. She spoke of its central character,
Harry the Mayor:

> he used to ride around on the donkey. He was the cowman
> at Heddon Steads. He used to have a bit of fun with all the
> young ones. He was always Harry the Mayor. There was a
> lot never knew what they called him.... They used to borrow
> the donkey and get him dressed up and he used to have for
> the Mayor's chain he maybe had a long dog-chain and a great
> big buttonhole. I remember one year it was the heart of a
> cabbage. He used to ride around the village on the donkey
> with all the children following behind. (NRO T/117)

After the picnic there was always a dance.
Dancing was a regular thing in Heddon. Mrs Hall explains:

> We had whist drives and dances in the reading room and the
> school. They got the place full. One played the piano and
> the other the fiddle.... There were always concerts at Hed-
> don; somebody would get a concert up. (NRO T/114)

Then there was the annual flower and vegetable show patronised
by the squire and Mr Bates. 'Everybody', says Mrs Hall, 'showed
stuff': 'and there was sport for the children. They used to have
foot races for the men, bicycle races for the men and a big

marquee. It was a calamity if it was a wet day'.

The social life of the village thus followed an annual cycle which governed such activities as growing vegetables, making new clothes, raising funds and the like. And just after Christmas, as Miss Elliot explains, the 'Guizers' used to tour the village to dance and to sing, soliciting drinks from door to door and playing mischief: 'At the new year you never knew what you were getting. They used ... you would hear them ... they used to roll the rain barrels down into the pond!' And on Christmas day itself the village was visited by the Throckley brass band and the Salvation Army band from Newburn. Since the village was closely connected with other villages up the valley the people of Heddon could easily join in the celebrations of others. Ovingham Goose Fair, for example, was a popular annual event even for the people of Heddon.

My grandfather particularly liked to go to the Stagshaw horse fair, only a few miles up the Military Road. He could keep in touch with horse prices and watch the rogues get rid of their nags. Sometimes he used to go to Hexham races for a day out or to Newcastle races. He went on occasion to a whippet race near Newcastle. There was no whippet racing in the immediate vicinity of Heddon although there was some greyhound coursing at Ovingham.

Beneath the organised life of the village, however, there was the illicit and the unorganised: pitch-and-toss, gambling, poaching, ferreting and thieving. Sarah Elliot spoke of poaching in the village. She once asked a friend of hers who lived at East Heddon how it was she was not afraid to travel home across the fields in the dark after the dance. She herself was terrified at such a prospect. But the friend replied nonchalantly, 'Oh, I'll just meet a few of the Heddon poachers or the Throckley ones!'

SETTLING DOWN

My grandfather rarely spoke about his youth although throughout his life he was deeply attached to Heddon. It is only possible, therefore, to convey something of the character of his days as a young man. He was tall and strong, and self-reliant as a matter of basic conviction. He was hardworking, healthy and well known in the village. Though even-tempered he was none the less quite ready to retaliate if he felt offended or cheated. He was tough, a little shy with women and gentle and indulgent with small children. He was clearly content with his lot. Only once that anyone is aware of did he consider leaving the pit. He used to tell of how he and a pal thought of joining the police force, but he decided against it because he was too honest! It is possible he considered joining the army. In the 1890s the army offered security and adventure in India, the Middle East and Southern Africa. It is more likely, however,

that what he looked for was a home of his own, free of the
congestion of Quarry Cottage.

My grandfather never spoke about girlfriends and he did not
marry until he was 27 years of age. The village clearly allowed
him many opportunities to meet girls – in the chapel, in the
reading room, at the village dance – and he could also travel
to neighbouring villages and to Newcastle. In fact, he travelled
to Newcastle most weekends to the market. I suspect, however,
that courting was not one of his great interests. He was busy
with his father's carrying business; he spent a great deal of
time at Law's farm and he liked a good drink with his mates.
The Brown family were seen in the village as mildly eccentric –
and I suspect, too, that because there were so many of them,
and that because the boys were all in the pits, and because they
were 'strangers' they were looked down on somewhat. All of
these factors might have prevented my grandfather feeling at
ease with girls; what is certain is that he married late and that
the girl he married was not from Heddon. She did, however,
come from an ordinary working-class family.

Born in Belmont, County Durham, a small village just outside
Durham City itself, she was the daughter of a pit sinker, a
worker who moved from one place to another wherever pits were
to be sunk. Although her family moved on to South Shields she
retained her sense of her Durham origins. She retained, too,
the Durham habit of dropping the letter 'h' from her words.

At the time she met my grandfather she worked as a shop-
keeper on the Scotswood Road in Newcastle (the road made
famous by the Geordie song, The Blaydon Races). They met in
Newcastle on a market day. My grandfather had taken the horse
and cart to town for the market; my grandmother was visiting
her sister who lived not far from Heddon and who knew my
grandfather. When they were introduced my grandfather was
slumped, quite drunk, across the neck of his horse. Pay Saturday – 'baff Saturday' – and market day combined was always a
good time to have a good drink. When he looked up through his
stupor he saw 'Aunt Maggie' standing with this young (she was
26 years old), tall stranger, and whether through devilment,
drunkenness or desire – no one knows – he proposed to her,
although in a pretty cack-handed fashion. 'Have ye browt me a
wife, then Maggie?' are the words he uttered. My grandmother
was mildly affronted by it, but obviously felt that this tall young
man with his cap and red neck-scarf had something about him.
My grandmother often told the story of how they first met. It
was always told, as it were 'against herself' as if to emphasise
that she had made a fatal error in marrying him. Within the year
they were married and within a year after that their first child,
Olive, was born. They were married in Heddon church and lived
first in a small, stone, terraced house just along from the Common where his parents lived.

Her education in Belmont had been much the same as his;
perhaps, if anything, a little worse. The log books of Belmont

school are just as preoccupied as those of Heddon with drumming
the three 'R's' into reluctant young heads and teaching them
the poetry of Longfellow and Mrs Hemans. The inspector's report
for 1883 notes: 'Girls' work very inaccurately done and attend
very badly'; only their needlework was well done. Belmont was
an old village, but during the second half of the nineteenth
century there grew up around it a pit and a steelworks, shops
and a chapel. Indeed, the chapel often offered alternative school
education free of charge, thus depleting the roll of the church
school my grandmother attended. I mention these things only to
underline the fact that my grandparents were of the same kind;
both were connected with pits; both knew small-village life; both
paid little heed to school and neither of them had any other
expectations than that they would marry, work and live much
as their parents had done. The only difference, is that they did
not want such a large family. My grandmother used to say - no
doubt partly with reference to her in-laws - that 'big families
are happy families; but they are poor ones'.

She gave up her work on getting married and settled into being a housewife, coping with the special problems of having a
husband in the pit. She was well prepared for this, though, as
were most young girls, and if the Belmont school books are to be
believed her domestic skills had been acquired at the expense of
her schooling.

She fitted into the Brown family with ease, visiting them
regularly and joking with the old man. She used to taunt him
with being a 'dirty old man' for having so many children, and
he used to reply, impishly, that 'them's not all I've got either';
but when she and my grandfather eventually moved to Throckley
she maintained a diplomatic distance from them such that her
own children did not get to know their grandparents very well
at all.

4 THROCKLEY

In 1900 my grandparents moved to Throckley. They moved first
to a house opposite the school - Gladstone House - and then,
shortly afterwards, to 177 Mount Pleasant, a two-bedroomed
terraced house owned by the Throckley coal company. They were
clearly determined to get into a pit house, and when Gladstone
House came up for sale they refused the offer of a loan so that
they themselves might buy it, preferring instead to await their
chance of a colliery house.

The colliery maintained a waiting list just as councils do now,
and when the Mount Pleasant house became vacant they leapt at
the chance to move in despite its poor decorative state and
general uncleanliness. Perhaps Gladstone House held painful
memories for them; their second child died there as a baby.
There was another incentive to move, too; my grandmother's
sister Maggie and her husband Harry lived in Mount Pleasant,
he being an official at the pit. But the overriding reason was a
search for security. My grandfather believed firmly that a colliery house was, as he used to say, 'a secure home'. The rent
and the coals were free and so long as his job was safe there
was nothing to worry about.

The notion of a 'free rent' and that of 'free coals' now seem
anachronistic since both were in reality part of his real wage.
But for a married man, during the period when sliding scales
were in operation (linking wages to the selling price of coal so
that both moved up and down together) free housing and heating
were stable components of his real income. Wages could go up
and down but his home and hearth would still be safe. The worries his parents had had over rent payments could be avoided
in a pit house.

His notion that his job would be secure is equally rational in
the circumstances. It was based on his own self-respect as a
pitman and a shrewd assessment, based on the contrast between
Throckley and Heddon pit, that the Throckley coal company was
a dynamic concern growing rapidly at this time and likely to
stay in business for a long time to come. He knew he was a good
worker, someone the company would value, and since he had no
ambition to move elsewhere, anticipating neither social mobility
nor even promotion in the pit, there was every reason to settle
for Throckley. Seen in this way the irrational thing to do would
be to burden himself with debt, thus threatening the independence which he valued so highly.

Throckley is two miles east of Heddon, built on the high slope

of the valley leading down to the Tyne with the west road from
Newcastle cutting it into two distinct halves. In 1851 there were
only 159 people in Throckley; by the turn of the century there
were more than 2,000. They were moving, therefore, into a
village that was growing very quickly, and for the first ten
years of their life there the growth continued. The figures for
the early part of the twentieth century are shown in Table 4.1.

Table 4.1 Population growth in Throckley, 1901-31

	No. of families	No. of people
1901	390	2,063
1911	530	2,612
1921	532	2,612
1931	589	2,640

Source: Census returns.

In 1851 the main facility of the village was a Methodist chapel.
By 1900 a church had been built (St Mary's, erected 1885-6),
a Wesleyan Methodist chapel (1870), a Primitive Methodist chapel
(1891), a school (1873), a mechanics' institute, a co-operative
store, several small shops, a church hall, a co-operative store
hall and, above all, row upon row of miners' houses named, in
sharp contrast to the unnatural underground world of the pit
workings over which they stood, after woodland trees - Pine
Street, Ash Street, Maple Street. Mount Pleasant, the name of
the street my grandfather lived in, was common to many colliery
villages. The street hardly lived up to its name, but at least it
had one: Ashington coal company further north gave its terraced rows only numbers. The houses closest to the pit they
called The Leazes with the connotation of pasture land, for that,
until the sinking of the pit in 1869, was what the place was.
Throckley, then, grew on and out of coal. But for coal Throckley
would not have existed at all except as a cluster of houses on
the turnpike road.

In this and the remaining chapters of this book I attempt to
describe what kind of community Throckley was, to show how it
changed during the course of my grandfather's life and to illustrate how the structures of community life in Throckley can be
understood through the patterns of everyday life of the people
who lived there.

My central point is this: Throckley was a constructed community with two historic impulses working through it. The first
of these, deriving from the actions of the coal company, was a
drive for capital accumulation and profit. As I shall illustrate,
this involved the coal company not just in capital investment
in the pit but also investment in a social infrastructure to attract,
support and control a mining labour force. Throckley from this

perspective was a creation of the coal company a community
designed to win coal from the ground which could be sold on a
market. Coal company policies, operating in the free market
environment of liberal capitalism, defined in a major way the
class position of the men they employed and the opportunities
inherent in that position. It was never just that to the coal
company; they were men of business but they also possessed an
image of themselves as social benefactors with responsibilities
to their employees extending far beyond the employment contract.

The second impulse is the rise of an organised labour movement and concerns the efforts of working people to gain a
greater control over their own lives, to guard against exploitation and to create institutions of their own to further their own
interests outside the control of the coal company which employed
them, and removed as far as possible from the vagaries of the
economic system in which they laboured. E.P. Thompson in an
often quoted passage pointed out: 'The working class made itself
as much as it was made' (1972, p. 213). I describe what it means
to say this in the specific case of the institutions which working
people in Throckley constructed for themselves.

Throckley was different from Heddon in that the village was
based, at least until the inter-war period in the twentieth century, almost entirely on coal, and the status, power and
authority of its ruling family was based entirely on industrial
capital. Throckley was a paternalistic village, almost a model
village, but it was the paternalism of industrial capital rather
than that of landownership which prevailed there. The early
vitality of the village, its rapid growth, its firm economic base
reflected the rising fortunes of a bourgeois class of which the
Stephensons and Spencers, the leading employers of the district,
were representative.

Throckley was not just a market for labour; the two impulses
I have briefly described fashioned the constraints within which
a community grew up possessed of values, relationships, attitudes and customs. For the people who lived there it was not
just a place on the turnpike road; it was a way of life with its
own distinctive nuances and symbolism, with particular commitments and obligations, and a framework of social relationships
in which people were known and could be recognised. Focusing
my account on the family life of my grandparents, I show what
was distinctive in that way of life.

The two historical impulses, or moments, I have briefly
sketched are part of one another. In the period before the First
World War they define a society of liberal capitalism and, in the
coalfield of Northumberland, of paternalistic capitalism. It was
not a stable order; while industrial co-operation and political
stability were both possible within it, change was its central
feature and conflict a frequent occurrence. And while the central
thrust of paternalism had always been, as E.P. Thompson (1972,
p. 222) argues, a managerial technique for coping with 'the
exploitive relationship' and the 'need for industrial peace, for

a stable labour force, and for a body of skilled experienced
workers', the period during which paternalism in the coalfield
was at its zenith is one which corresponds with a shift in the
consciousness of ordinary workers which allowed them to penetrate its façades and conceive of a new kind of industrial order.
I seek to show how this change occurred in my grandfather's
awareness of the world around him and, through him, to show
how working people acquired new meanings for their experience.

When my grandparents moved to Throckley Britain was still
fighting a colonial war in South Africa. I do not really know what
he thought about that. He once told me, as I struggled with a
history essay on Southern Africa, that 'Kruger was an owld
bugga'. I suspect, given his schooling, he felt that national
pride was somehow at stake, a view heavily endorsed by the
press and even the pulpit. I mention this for a very specific
reason: to introduce the Throckley coal company. Major Stephenson, the nephew of Sir William Stephenson one of the Throckley
coal-owners, had fought in the South African war with the
Elswick Battalion of the Royal Northumberland Fusiliers. He
returned to the village shortly after my grandparents had settled in and that return was an excellent introduction to the
social position of the Stephenson family in Throckley. Major
Stephenson's return to a hero's welcome has acquired something
of the status of the folk myth in the district and my grandfather often referred back to it with a tellingly indulgent smile.

The train from Newcastle stopped at Newburn and there was a
hugh crowd waiting to meet him. His horse and trap were ready
for him but the horse, as things turned out, was not needed.
A group of miners unhitched the horse and pulled the major all
the way up Newburn Road to Throckley. And stretched across
Coach Road, the road down to Throckley House, there was a
banner which read 'Welcome Home, Willy, We've Missed You'.
Major Stephenson (my grandfather, like many of the older people
of Throckley, pronounced it 'Stivvyson') was the heir apparent
to the coal company, a local politician, a stalwart of the Royal
Northumberland Volunteers and a man much respected in the
village. How he acquired that position can only be explained if
the history of the coal company itself is examined. What follows,
therefore, is an account of the development of the company and
how it built Throckley. G.M. Norris (1978, p. 469) has suggested that systematic work on industrial paternalism has yet to
be carried out. This chapter, and the following one, is a small
contribution.

THE THROCKLEY COAL COMPANY

The Throckley coal company was set up in 1867 to work the
Throckley royalty, then owned by the Greenwich Hospital and
the Lords of the Admiralty, on ground formerly owned by the
earl of Derwentwater. Two local families were its principal

shareholders, the Stephensons and the Spencers, the first being
brick manufacturers and farmers who had leased farms in the
area since 1782, the second being steel manufacturers from
Newburn owning at that time one of the largest steelworks in the
north of England. Two other shareholders were J.B. Simpson
and E.J. Boyd, both mining engineers. The shares at this stage
had a nominal value of £1,000 each. They anticipated a capital
requirement of £18,000-20,000 to develop the colliery and, with
the agreement among themselves to float the company with the
share capital just described, moves were quickly set in motion
to build workshops, railways and to sink the shaft. On 24 April
1867 Isabella Stephenson cut the first sod and the pit was named
after her, known locally as the 'Isabella'.

The men who formed the company were very respectable local
businessmen. Coal had been worked extensively in the district
for centuries, but the small local pit - the Bobby pit - which
supplied the Stephensons' brickworks (the largest in the region)
with both its fire-clay and its coal was not big enough or deep
enough to produce coal on the scale the company required. More
efficient pumping engines opened up the possibility that the
lower seams in the Throckley area could be won, and this,
together with the firm local market for coal which the Spencer's
steelworks guaranteed, made the sinking of a new pit at Throckley a commercially sound enterprise.

The development of the company illustrates clearly that process of industrial growth which economists describe as vertical
integration. It is a theme to which I shall return since, having
a firm local market for its output, although not exclusively
supplying to local industry, the Throckley coal company was
saved from many of the difficulties of export price fluctuation
which affected so many pits in the Northumberland coalfield.
This immunity from export market pressures is a significant
factor in explaining the relatively peaceful character of industrial relations in Throckley. Initially, however, a firm local
market was what justified the new pit.

The prominence of the local businessmen and the Throckley
coal owners was acknowledged throughout industrial Tyneside
for their business connections extended all over the region. They
were civic benefactors, too; Sir William Stephenson (knighted
in 1900) donated public libraries to the City of Newcastle, and
J.B. Simpson gave substantial sums of money to higher education in the area. E.J. Boyd became high sheriff of Durham and
they each, at different times, held high public offices. Sir
William Stephenson was mayor of Newcastle (see Williamson,
1980).

Judged as businessmen they were clearly very successful and
enterprising and their business interests expanded rapidly
throughout the latter part of Victoria's reign. They diversified
their investment from mining into public utilities, engineering
and shipping and their public prominence rose as a consequence.
The Stephensons and the Simpsons, however, never lost sight

of their commitment to Throckley itself. Their view of themselves as public benefactors who had the interests of their
employees at heart was a key element in the way they established
all the social preconditions of a successful mining company.

Apart from capital to invest, mining companies need miners
and miners need houses. Out of the initial capital of about
£20,000 the company calculated that £8,000 would be needed for
housing alone. In the 1870s the provision of decent housing
was an important factor in being able to recruit miners (see
Daunton, 1979b). By the standards of the day the houses provided in Throckley were good. A 'Newcastle Weekly Chronicle'
report in the 1870s, part of a series of articles on mining villages, noted that company houses in Throckley were 'the best
kind as yet erected for pitmen and all have all the proper con-
veniences' (16 November 1872, p. 4). Sewerage facilities were
none the less of a pretty rudimentary character with traceable
effects on health.

The company managed its houses carefully; they did not permit
tenants to keep dogs - because the Stephensons wished to protect their shooting rights - and they were careful to remove
unsatisfactory tenants. Allotments were available to virtually all
who wanted them and the company encouraged their men to keep
gardens. Miners who were sacked and the widows of miners
killed had, however, to vacate their homes. In this way the
control of housing was an essential element in the control of
workers, but in the Throckley case it was a control exercised
benignly and I could find no evidence that housing created
special difficulties. Indeed, J.B. Simpson provided from his own
pocket a whole row of cottages for aged miners, an act seen as
one of great generosity at the time.

Throckley was not just houses. As leading Methodists in the
Newcastle West circuit, the Stephenson family had built a Wesleyan chapel and a school in the village. The Stephensons helped
to build chapels in Newburn and Blucher, neighbouring villages,
and Sir William was a leading member of the Newcastle School
Board. The coal company constituted itself the management
committee of the colliery school and they were closely involved
in the running of the school. When some of their workmen failed
to pay school pence they simply docked the charge from their
wages.

The company's presence at the centre of social life in Throckley
was exemplified in many ways. They donated money to the pit
brass band. Major Stephenson occasionally played for the local
football team. Matt Cheesman, the pit under-manager, was a
keen pigeon fancier. Major Stephenson sometimes took some of
his employees shooting. Major Stephenson's wife went sick visiting and took a special interest in visiting women with small
babies to give them advice on child rearing. These are clear
aspects of a pervasive ideology of localism which was such a
major part of late Victorian paternalism. I found it virtually
impossible to press any of the people I talked to in Throckley to

say anything against the coal company as an employer. I was
told time and time again that they were a good company to work
for. Certainly my grandfather held this view. He said often
enough that Major Stephenson was a gentleman. The target for
my grandfather's bitterness, such as it was, for he was not a
bitter man, was not the company at all but the industry as a
whole. For as an employer of labour the company was bound by
the agreements between the Northumberland Coal Owners' Association and the Northumberland Miners. These agreements bound
them from 1879 onwards to a framework of sliding scales and
joint committees to solve industrial relations questions which
were unique to the Northumberland coalfield. What this meant,
of course, was this: any conflict about price changes or wage
levels could be deflected as concerning the coalfield as a whole
and not being in any way specific to Throckley itself.

The company operated with a view to the identity of interests
of workmen and employers and were strong supporters of the
sliding scale method of payment. J.B. Simpson, in a lecture to
the Newcastle Economic Society in 1898, argued that such scales
would 'enable us to obliterate from our commercial dictionaries
the terms "strike and lockout"' (Simpson, 1900, p. 48). They
were not sentimental about the relationship, however. Charlton
Thompson told me he once overheard a workman asking a special
favour of Major Stephenson, prefacing his comments with the
remark, 'I have worked well for you for over thirty-years.'
The major replied before hearing the request, 'But we've paid
you, haven't we?'

The company was considered to be a good one to work for,
particularly before the First World War. This was partly due,
of course, to its commercial success. I calculate that up to the
end of the century the company was making a 15 per cent return
on capital invested (Williamson, 1980). By the end of the century they had acquired Heddon and Blucher collieries. In 1898
they employed 724 men. By 1904 this figure had doubled to
1,484. Since their production was not geared entirely to the
export market, the Throckley pits were not subject to the same
economic stresses as pits elsewhere in the coalfield. Wages were
stable over longer periods and this encouraged good labour
relations. How labour relations were handled underground is
something which I discuss in the following chapter.

For the moment, however, it is sufficient to note that the
company was paternalistic with many formal and informal contacts
with the whole associational life of the village. The Stephensons
were known in the village; the 'big house' in which they lived
was not inaccessible. As employers they were not distant figures
or part of a larger coal combine, and their contact with the
community had, I believe, important consequences for the way
in which the Stephensons were perceived by people in the village. Personal contact and noblesse oblige had the effect, in this
most traditional of industries, of obscuring the argument for
many Throckley men that the cash nexus and naked self-interest

were the main preoccupations of capitalist employers. Certainly
my grandfather did not regard his employers in these terms,
and given how they did relate to the village and to men like
himself, there was little in his direct experience of his employers
to convince him otherwise. He was not deferential towards them
and he did not share their political views, but he did respect
them and continued to do so for the whole of his working life.
The comparison he made was not between the Stephensons and
some nationalised enterprise of the future, but between Throckley pit and other pits in the district. Judged by those standards
it was a good company to work for just as long as 'everybody
played the game' properly.

COMMUNITY: STRUCTURE AND DIVISION

It is a central theme of this book that the mining village is a
constructed community. In the remainder of this chapter I want
to examine briefly some of the institutions of the village which
were constructed by the miners themselves and which were
central to the life of the community as a whole. It is quite impossible, however, to disentangle cleanly that which reflected the
action of the coal company and that which evolved from the men.
It is also impossible to distinguish those institutions which are
or were uniquely of the village itself. As I have explained in
the Introduction, the model of the mining community as an
isolated community is entirely misleading. The image of the mining community as the 'archetypal proletarian community' (Bulmer,
1975) is equally misleading. The social mosaic of life in Throckley
reflected a great diversity among the people who lived there.
And it is essential to understand that diversity and the structures which produced it if the attitudes and actions of my grandparents are to be explained.
    Settling down in Throckley meant to them coming to know and
participate in (although selectively) the associational life of the
village, a life built around the major institutions of the village.
In addition to the brass band subsidised by the coal company,
there was a football team, Throckley Villa, and a cricket team.
The mechanics' institute (this, too, having been provided by
the coal company) had books and a billiard table. As early as
1872 the 'Weekly Chronicle' had noted that the institute possessed
a 'pretty fair library' and was 'modestly supplied with papers'.
The same report, however, detected the lack in the village of a
proper working men's club and noted that the fifty members of
the institute looked forward to having such a facility. The paper
adds: 'The way to these blessings do not seem just now very
clear; but the men of Throckley are hopeful, cheerful, patient
and persevering, or they would not be such good miners as they
are' ('Newcastle Weekly Chronicle', 16 November 1872). They
had to wait, in fact, another thirty-five years before that
particular aspiration was realised.

In 1905 the Throckley Rechabites - the 'Pride of Throckley'
Tent of the Independent Order of Rechabites, the main temperance body of the village - boasted a membership of 588 people;
208 adult members, 28 members' wives, a female tent of 32 and
a juvenile tent of 320. Mr M. Heslop ran a Home Reading Union
circle and William King a cycle club which, apart from planning
outings, raised money for charity. Local pubs in nearby villages
ran annual leek and vegetable shows offering to the winners
modestly practical prizes such as blankets. The union ran social
events, too. In 1905, for instance, the union organised a
'novelty football match' in aid of the Sick and Accident Fund.
The colliery band paraded around the village and there was a
public tea in the store hall in the evening.

The church and the chapels were at the centre of other organised activities. Attached to the church there was a branch of
the Mothers' Union. There were Sunday schools. Regular evenings of dancing were held at the chapels which were important
social centres where boys and girls met. A fair visited the
village every Easter - the 'hoppings' - and there were occasional
concerts given by local choirs and brass bands. For the men
there were pigeon-fancying clubs, and whippet-racing meetings
were held along the back lonnen behind the pit. Woven into the
leisure time of Throckley families these local groups made their
own contribution to the flow of everyday life, reinforcing the
prevailing sense of belonging to a particular place and a particular class. To come from Throckley meant precisely this - to
be recognised as someone who shared in the associational life of
the village. How my grandparents participated is something I
discuss in subsequent chapters.

SOCIAL CONDITIONS

Social conditions in the village were, however, very poor; the
reports of the medical officer of health for the district are
remarkable. They show a village with a high infant mortality
rate, poor sanitation, inadequate housing and overcrowding.
The fundamental problem was housing; poor housing, he felt,
lay behind the high incidence of preventable disease and infant
mortality. He wrote in his 1906 report: 'The health of the community is now largely in the hands of the capitalist and the
architect, and as Medical Officer of Health I can only lay down
general hygiene principles' (Medical Officer of Health, Newburn
UDC). Of the architecture he had this to say:

There are in every part of the district beams and brickbats
thrown together in the shape of houses. Naturally they are
damp, many ill lighted, badly ventilated ... which go to make
them unfit for habitation. As for the tenants

Alas their sorrows in their bosoms dwell
They've much to suffer but
have nought to tell

This was written in 1906. In 1910 his report was much concerned with overcrowding: 'The housing accommodation for this
large population is one of the least satisfactory features of the
district. Eighty-four per cent of the people were lodged in
houses of less than five rooms at the last census.' And in this
same report the medical officer of health sets out his diagnosis:

The cause of it all? Not ignorance, we know better; and the
builder is not to blame, for the law allows it. It is disheartening at times to feel, that after all these tons of reports of
Royal Commission, Medical Officers and Sanitarians, cartloads
of books and pamphlets which have been written since the
days of Cobbett to the present time, we still go on in the
same way. Possibly Tennyson's old farmer would make his
pony give a shrewd guess at the reason why we do not progress.

'Proputty, proputty, proputty, that's
what I 'ears them saay.'

In 1912 he sounded a note of warning about the recent budget
from the chancellor of the exchequer:

The increase of wealth as shown by the Budget is in the
large fortunes of the few, rather than in the diffusion of
nobler possibilities of happiness among the masses. In spite
of the booming trade the vast mass are no better, and there
is always the liability to forget this under the hypnotic
influence of big figures.

Dr Messer's reports to the health committee played an important
role in fashioning the outlook of local labour councillors and
the man himself, for a while after the First World War, became
a Labour member of Northumberland County Council. They are
remarkable reports, however, because they cut through a great
deal of the complacency surrounding public welfare in the area
and they locate the root causes of poor housing and high infant
mortality in the capitalist system itself.

The terrible irony must be noted that the conditions which
Dr Messer describes reappear in the lower echelons of the subtle
structures of hierarchy in Throckley which allowed people to
distinguish themselves from one another. The obvious dangers
to the poorest people in conditions such as these (Dr Messer
commented in his 1917 report, 'the state of overcrowding in
many houses is deplorable, a terrible menace to moral and
physical well-being') were simultaneously an opportunity for
others to prove their respectability. In 1916 the health visitor

for the district, following a survey of 1,444 homes, reported to
the Health Committee in the following way:

> A small percentage of homes visited were found to be very
> dirty, a disgrace to the housewife and a peril to the inmates
> and to the neighbours. On the other hand, many houses are
> always beautifully clean, making one realise that the real
> heroines of England are the mothers who, day by day, are
> faithfully fighting dirt and disease sometimes under heavy
> odds. (Medical Officer of Health, Newburn UDC)

Overcrowding was one serious problem. Another was the threat
of loss of earnings. That threat was poverty. I have no data for
Throckley on the extent of poverty but Throckley was covered
under the Poor Law by the Castle Ward Union, so there is
evidence of official attitudes towards it. The Castle Ward Union
in 1901 covered 32,357 people of whom 12,500 were in the Newburn area ('Royal Commission on the Poor Laws and Relief of
Distress', Appendix no. CXVI, vol. V, 1909). Outdoor relief
in 1901 was paid at a rate of 2s. 6d. for adults and 1s. 6d. for
children. For the 'really destitute old people' it rose to 3s.
  In his statement of evidence to the Royal Commission on the
Poor Laws Canon Walker of Whalton, secretary of the Northern
Poor Law Conference, describing the policy of the Castle Ward
Union, pointed out: 'The proportion of paupers to population
is little more than 1.3 per cent, and this is found mostly in the
industrial areas where the higher wages prevail with high rents.'
He went on to note: 'The classes applying for relief are mostly
labourers who have been employed about the pits or factories,
disabled by age or infirmity; and young widows with children.'
These were the precise categories into which all the people of
Throckley could so easily have fitted but for their own good
luck and the care they took with their work.
  Canon Walker touched on issues which preoccupied Dr Messer,
the medical officer of health. But the two men did, in fact,
arrive at quite different explanations of poverty, and the good
canon gives us some clue to the way in which Poor Law guardians
were encouraged to view the plight of claimants:

> The causes of pauperism are extravagance, money spent on
> dress, in pleasure and amusements; improvident habits,
> especially amongst the girls and young women and a feeling
> often expressed that there is a right to Poor Law Relief;
> improvident marriages, phthisis and it may be high rents.
> The housing question has a serious bearing on pauperism.

This curious passage moves from explanations of pauperism
based on the moral worthlessness of the individual to those centred on poor social and occupational conditions. That the latter
might explain the former is not something which occurs to the
canon. He does point out, however, that the guardians are not

severe people:

> We do not try experiments in this union. The majority of the
> guardians are rather inclined to give out relief in preference
> to in-maintenance. It is only in the case of old and infirm
> people, or where a man is a well-known loafer or waistrel that
> an order to the house is given.

But he is not himself wholly in agreement with this policy
administered as it is by miners, colliery managers and retired
tradesmen, 'men', he says 'who are well acquainted with the
people of their districts and are all the best of their class, men
of good judgement'. 'Even the miners', he says, as if anyone
might doubt it, 'who represent their unions are considerate,
prudent, and desirous of administering the Poor Law with
equity.' The canon himself, however, was more in favour of the
workhouse and the only constraint on him was lack of space:
'Personally, I advocate a more frequent use of an "order of the
house", but we could not exercise it with any freedom except at
great expenditure on buildings.' Even so, when we turn to
census reports we find that in 1911 there were 3,374 people in
workhouses in Northumberland, 96 of whom were in the Castle
Ward Union. Even as late as 1931 the figures were 1,411 and 54
respectively.

Right through my grandfather's working life, although less
so near the end of it, he felt the threat of the workhouse as
ever-present if misfortune were to overtake him. His mother was
born in a workhouse. Would he die in one? And my grandmother
did really worry about it, communicating a fear to her children
which they have never forgotten.

Social conditions and social attitudes of the sort outlined are
simultaneously constraints on and conditions for social change.
They represent a field for action, both for the individual and
for the community as a whole. For the individual they prompt
efforts at self-improvement, self-protection and change; for the
community they prompt neighbourly self-help and political and
industrial action.

SOCIAL DIFFERENTIATION: PERCEPTION AND PARADOX

It would be too simplistic to say that such experiences and feelings give rise to a uniformity of political viewpoint. There were
clear political divisions in the village reflecting cleavages in the
labour movement and social differentiation in Throckley itself.
Throckley was not a homogeneous community. There were several
social fissures despite the fact that most of the families were
connected with the pit. There were, firstly, territorial divisions.
For instance, the people who lived north of the West Road in
the so-called High Rows leading to the Maria pit saw themselves
as a separate group from those in Mount Pleasant and The Leazes.

These divisions generated a mild social rivalry, particularly for young people, which found expression in football matches and, among the boys, in what Albert Matthewson calls 'sham fights' with poles! After the First World War new territorial divisions emerged along with the house-building programme.

Alongside these were the religious differences. Church people and chapel people are said to have been quite distinct and to have kept themselves apart from one another. Even within the nonconformist group there was a division between the Wesleyans and the Primitives. These divisions had no great political importance and for most practical purposes did not matter at all. Indeed, my family were always connected with the church but my mother and her brothers and sisters joined in with the chapel people in whist drives, dances and outings, there being no personal animosities involved.

Class was a potent principle. The barriers which it implied circumscribed completely how people *felt* their social position. C.F.G. Masterman, the great Liberal MP and social reformer of Edwardian England, commented on class in the following way, revealing the great chasm which separated the different classes:

> We are gradually learning that 'the people of England' are as different from, and as unknown to, the classes that investigate, observe and record, as the people of China and Peru. Living amongst us, never becoming articulate, finding even in their directly elected representatives types remote from their own, these people grow and flourish and die with their own codes of humour, their special beliefs and moralities, their judgement and their condemnation of the classes to whom has been given leisure and material advantage. (Masterman, 1911, p. 98)

And he goes on to quote one observer commenting on the divisions between the classes: 'There is not one high wall but two high walls between the classes and the masses and that erected in self defence by the exploited is the higher and more difficult to climb' (ibid., p. 99). How far these divisions were breached is a profound measure of social change. For the moment I want only to add that there were strong and subtle differences among workers themselves.

Throckley was dominated by the big house and the Stephenson family, and the pit itself produced other divisions. The gap between the pitmen and the pit officials was an important one. Officials not only had more money and power, they had better houses and closer contacts with the Stephensons. But even among workmen themselves there were the subtle divisions between hewers and datal men, underground workers and surface workers, differences which translated themselves not only into different wages but also into social status. There were also divisions between miners and other groups of workers. Albert Matthewson told me that some of the steel men from Newburn who were earning good money at Spencer's thought the miners 'beneath contempt'.

Overriding the differences among the miners themselves and other working men there was, of course, the bigger gulf between the miners and the local white-collar workers of various kinds. The traditional professionals the vicar, the minister, the doctor - were, of course, accorded a great deal of respect. This was not so true of the shopkeepers or the clerical workers employed by the coal company, the store or the urban district council. A typical reaction to such people among miners was to deny them their basic masculinity. Several people have reported to me a form of abuse which goes: 'He couldn't hew himself out of a paper bag.' This was always deployed against ostentatious shows of clerical superiority or arrogance. My grandfather's favourite comment on such cases was: 'If they paid him his wages in coppers he wouldn't be strong enough to carry them.' Given their education pitmen could hardly make a claim to status on grounds of knowledge; they could have claimed a greater right to respect on the basis of their underground skills. But since in Throckley these skills were not regarded as special the only recourse was to stress their toughness.

In the competition for the hallmarks of status, miners in Throckley were seriously handicapped because they had so much in common; there were few ways in which they could easily distinguish themselves from one another. Mark Benney made much of this fact and drew from it an important conclusion:

> The income of every mining family, insofar as it is derived from the mine, was known to everybody. The pretensions of urban living were impossible here. No family could assume higher standards than its income warranted without incur- ring ridicule. Here, perhaps, part of the reason why miners made their demands on life as a community, not as individuals. (Benney, 1946, p. 24)

Paradoxically, perhaps, it was because the inhabitants acted as a community that another division appeared in Throckley, that between natives of the place and 'outsiders'. It was always evident, of course; but in the 1930s, as I shall show, it coloured the policy of the union lodges on questions of employment and unemployment. They sought to give men from Throckley priority in all matters to do with the pit. In certain circumstances, locality was a far more potent symbol of belonging than class.

That there were few differences among Throckley miners in respect of social and economic position had another almost para- doxical effect. Claims to be different had to be based on more abstract qualities such as respectability, bearing, honesty, integrity, learning.

Among women the criteria which operated involved such notions as propriety, i.e. whether or not a woman moved outside the strict limits, both in her work and leisure, which defined the

respectable wife. Tidiness, cleanliness, keeping children under
control were also of great significance. Women who let their
houses become dirty or did not see to their children were looked
down on. Albert Matthewson, reflecting on the difficulties faced
by women in Throckley ('I don't know how they survived'), told
me that 'some just didn't care. Their houses were like middens'

Some families came to be labelled as particularly scruffy or
unrespectable. They were not cut off socially but they represented a kind of benchmark against which the social status of
the rest could be measured.

The result was this: since it was assumed that everyone was
in the same boat, any evidence of falling behind, of slovenliness
or of poor standards of work - any failure, in fact, to meet the
current standards of respectability - was taken as an indictment
of the individual concerned and not as a reflection of prevailing
social conditions. Among other effects, perhaps, the main one
was to stigmatise those who would demean themselves by going
to the 'guardians' for financial help and to look down on those
who persistently borrowed or scrounged.

The same sense that much was held in common operated to
censure those who sought to stand above everybody else and
claim social distinction. Whenever this happened the person in
question would immediately be reminded of his past and questioned aggressively with 'And who do you think *ye* are?' An
upward change in social status which did not incur the contempt
of others was only possible if it involved an explicit acknow-
ledgment by those affected that they were still part of the community and in many other respects were just the same as anybody else. Here, perhaps, is the clue to why those with sufficient detachment from the village, with some ability to see its
faults and sense its possibilities for improvement, sought office
in the union or, after the First World War, in the Labour Party
or the co-operative store rather than in personal aggrandise-
ment. Public political life brought the rewards of status and
the respect of others and avoided the acrimony of individual
social climbing. As Mark Benney says, they made their demands
as a community and not as individuals.

POLITICAL PARTIES

The political mood of the village at the turn of the century, and
of the Wansbeck divisional constituency of which it was a part,
was Liberal. Charles Fenwick, Thomas Burt's good friend and
parliamentary colleague, a miner himself, was the MP. The
active political force in the village was the Throckley-Walbottle
Liberal Association, presided over in 1905 by Councillor Kirton.
James Bestford was secretary and Thomas King the treasurer,
all of them leading co-operators. There was a small local branch
of the Conservative and Unionist Party with political representation on the urban district council. In 1914 its president was

Major Stephenson, the managing director of the coal company. However, the active political group, at least until 1905-6, was the Liberals, the party of Gladstone, free trade, home rule for Ireland and social reform.

Some of their political preoccupations are apparent in local newspaper reports of their meetings in 1905 and 1906, the time of the great Liberal election victory. The 'Blaydon Courier' reported a meeting of the branch in 1905, when it was addressed by a Mr Vietch of Newcastle and the parliamentary candidate for the Tynedale division, Mr. J.M. Robertson. It noted the former's comments that the present government (i.e. of Mr Balfour) was 'a menace to good government and the dignity and efficiency of the House of Commons':

> The influence of the private member in the House of Commons had been reduced to nil. The government did not like men to express their own minds but wanted them to follow simply as hacks at the heels of Mr. Balfour.

Mr Vietch went on to hope that the next Liberal government would reduce taxation and have clearly thought-out policies. The meeting was then addressed by Mr Robertson. His main point was: 'It was the business of a government of a great industrial country to take scientific measures for dealing with unemployment as a more or less permanent fact.' His own solution involved measures such as technical education and public works and more effective taxation of landlords. This particular remark has a clear radical thrust and he followed it with an appeal to the labour question:

> The test that a practical working man would put to any party was: What are they going to do for the betterment of the condition of labour? Any government in this country ought to be prepared to stand or fall by that test. ('Blaydon Courier', 21 October 1905)

Robinson was successful in the 1906 general election and as an MP addressed the Throckley Liberals in the October. His theme this time was to distinguish the Liberals from the Labour Party. In retrospect his address was a timely one for 1906 was a turning point; the Liberals were at their zenith but poised for a long-term demise, undermined by the growing Labour Party formed in that year.

> There were more socialists in the Liberal Party than in the Labour Party - that was, men who believed that the evolution of society was in a Socialistic direction. But these men were what might be called philosophical Socialists and were not disposed to make attacks on property or confiscate wealth. They were Liberals who saw ahead. (Applause) ('Blaydon Courier', 13 October 1906)

There were, however, other voices to be heard. This was a
period of great activity for the Independent Labour Party (ILP)
and the movement established many branches throughout the
Northumberland and Durham coalfield (Gregory, 1968).

In April 1905 there were only four ILP branches in Northumberland. These were in Newcastle, North Shields, Ashington and
Wallsend (Purdue, 1974). By 1906 there were branches at Bed-
lington, Benwell, Pegswood, Blyth, Walker, Seaton Delaval,
Cramlington and Throckley. In the Throckley branch the active
members were Richard (Dick or Dickie) Browell from Blucher,
Dan Dawson, J.H. ('Henna') Brown and George Curwen from
Throckley. They were all miners and later became officials in
their lodges. Browell and Dawson were staunch Methodists and
they maintained a very active branch. Mrs Gibb of Morpeth,
an old Labour stalwart and north regional organiser of the
Labour Party in the inter-war period, told me that the meetings
they organised were 'terrifically well attended'. Dan Dawson,
she said, was 'politically obsessed'. He was a very effective
organiser and had the power to make branch members feel
obliged to attend meetings. The ILP, she told me, was effective
among ordinary miners both in Throckley and district and
throughout the coalfield. The Throckley branch meetings in the
co-operative store hall were held on Sunday evenings and they
were packed. 'I can recall the atmosphere at that packed co-op
hall. It was like a well-attended church.'

The Throckley ILP continued until the 1930s; it was this
organisation, in fact, which gave precise definition to the Labour
politics of the Throckley activists; when, after the First World
War, Labour Parties were formed in the area (i.e. in the Wansbeck division and the Newburn district) the principal officers
were ILP men. The ILP directly shaped the character of the
Labour Party in the area and was active in the union lodges.

The ILP campaigned in this period for Labour representation
in Parliament, a demand which brought them into conflict with
the Liberals. The tone of their political engagement with the
Liberals can be gauged from a report of an address to an ILP
meeting given by Philip Snowden in Benwell, four miles east
of Throckley near Newcastle. His address concerned unemployment:

> They had had something like 80 years of Liberalism and
> Toryism, and what was the position of the working people
> of the country to-day? They had nothing to conserve to be
> liberal with, and therefore, they owed no gratitude to either
> party for anything that had in the past been done for them.

He went on:

> The agitation on behalf of the unemployed had been carried
> on exclusively by the Socialists and the Labour Party in this
> country (applause).... All the industrial and social reforms

secured during the last thirty or forty years had been won
from the Liberals and Tory Governments by Trades Unions.
(Applause) ('Blaydon Courier', 21 October 1905)

He finished with a savage criticism of the House of Lords, and
after his speech the meeting ended with songs from a local choir
and a violin solo.

The rise of the ILP brought conflict, too, within the union
which directly threatened the position of Thomas Burt and
Charles Fenwick, the ageing leaders of the miners (Gregory,
1968). The union leadership tried to portray the ILP as 'outsiders' whose influence was resented (NMA Executive Minutes,
1908, NRO 759/68). The ILP response, articulated by Philip
Snowden at the Northumberland Miners' picnic in 1908, was:
'The differences which had existed between trades unionists and
socialists were very small. In fact the trades unionists had been
socialist all the time without knowing it' (ibid.). The union
decided finally in 1911 to sponsor only Labour candidates. Under
the leadership of William Straker, Burt's successor as general
secretary, this change was completed. The union had, therefore, between 1906 and 1911 switched its allegiance from the
Liberals to Labour. How this shift was experienced in the lodges
is a question yet to be researched. It is likely that older miners
never lost their attachment to the Lib-Lab politics of Burt and
Fenwick. 'Ye monna say nowt agyen Burt and Fenwick' was the
advice given to Matt Simm, the ILP activist, when he launched
their monthly newspaper, the 'Northern Democrat' (Gregory,
1968, p. 74).

In the Throckley case, however, the socialists were strong
and effective but the Liberals were, too. My uncle Bill insists
that during this period my grandfather clung to the older traditions and even helped the Liberals at election times. For him it
was still the party of the working man. Liberalism forms an
important link with another key institution of the village, the
co-operative store.

THE CO-OPERATIVE STORE

When my grandparents moved to Throckley one of their first
acts was to join the co-operative store. Going to work brought
my grandfather into contact with the union; feeding the family
introduced him to the store. That the meaning and purpose of
particular institutions change over time, reflecting larger shifts
in the political temper of the working class as a whole, is a
theme very well illustrated in the case of this institution.

G.D.H. Cole notes in his history of the co-operative movement that in the last quarter of the nineteenth century, a period
of very rapid growth, the attachment of the movement to free
trade ensured that

the effective weight of the Movement was as much on the
side of Liberalism as was the weight of Nonconformity. Religious Dissent and Consumer's Co-operation were twin props
of Liberalism; and the young men who had embraced Socialism
and were struggling to convert the working class movement
to the Socialist faith found the going very much heavier in
the Co-operative societies than in the Trades Unions. (Cole,
1945, p. 192)

The movement, he says, 'had not passed through the same
evolutionary experience as the Trades Unions'. Unlike them, it
was not involved in direct class struggle; it was more concerned
with consumers' interests, and in the period of expansion both
of unions (particularly for the unskilled) and of co-operation
they drifted further apart, the earlier connection which had
existed between them and which had been based on Owenite
Socialism and Christian Socialism becoming a mere dull echo of
a receding past.
Between 1881 and 1900, the period in which Throckley store
was founded, national membership of retailing societies increased
from 547,000 to 1,707,000 (Cole, 1945, p. 212). In two important
ways the Throckley store was typical of the movement in this
period: its executive committee was entirely male and its active
leaders were also active in the local Liberal Association. This
situation did not alter substantially until after the First World
War. In both respects the character of the store reflected not
just the position of the movement but the attitude of late Victorian society as a whole to the position of women and the state
of working-class politics. Cole says of the former that 'it was
still a very common notion in all classes that women were unfit-
ted for the conduct of business and ought not to take part in
public discussion' (ibid., p. 184).
'The stores', as it was and still is referred to, was founded in
1887 and became an independent society in 1892. Prior to that
people in Throckley had to shop either in the small shop leased
to Mr Henderson by the coal company or travel to Lemington,
Newburn or even over the river to Blaydon to shop at the co-
operative stores there. The bi-centenary brochure of the society
puts it glowingly:

> After years of weary trudging down the COLLIERY WAGON-
> WAY to LEMINGTON for their groceries and provisions –
> some of whom, we are told, took with them handbarrows –
> an opinion was apparent that their PARENT SOCIETY should
> provide shopping facilities in their own district. This was
> ultimately achieved in 1887. Having now tasted the fruits of
> actual shopping facilities and realised the benefits of co-
> operation in their own *little village*, it was not long before
> the IDEALISTS, imbued with the spirit of co-operation and
> dreams of self preservation, visualised the possibility of a
> VILLAGE STORE of their own. (Throckley District Co-

operative Society, Souvenir Report and Balance Sheet,
1942)

As the main commercial retail store it gained the custom of
nearly everyone in the village. But from the beginning it offered
other services. The tailoring department, no doubt in recognition of the fact that it was serving a mining community, used
to advertise itself, for instance, with the slogan, 'Mourning
orders executed on the shortest notice'. It offered a burial
service, an insurance service and, above all, it acted as a
building society and even provided houses itself.

Before the turn of the century it had built sixty-three houses
- in Hilda Terrace, Orchard Terrace and Victoria Terrace - and
sold them to co-operative members at cost price. Such houses
were thought of in the village as being superior to colliery
houses and even now they are still in good repair, many of the
colliery houses having been demolished. In 1910 they built a
further nineteen houses at Throckley, allocating them by ballot
among those members who had applied for them. For the moment,
however, the point is this: the hegemony of the coal company
was breached considerably by the co-operative store, and during the inter-war period, as I shall show later, particularly during the troubles of 1921 and 1926, it performed a vital function
in extending credit to families without any income at all.

The meetings of the Co-operative Society were well attended
and very lively. Mr Stobbart, a well-known centenarian in
Throckley, told me that the quarterly meetings 'were better
than a Saturday night out. Everybody wanted their say to argue
with the committee'. The reason, of course, was that the store
controlled a great many resources and job opportunities and was
one of the few institutions in the village (apart from the union)
which was properly accountable to ordinary people themselves,
an accountability which they were reluctant to see slip through
their hands.

Membership of the store was easily achieved although it was
confined to men. The initial cost was £5, the price of five £1
shares in the business, which members could either pay at
once or, as most of them did, in instalments. The benefits of
membership were wide. Profits were returned to members as
quarterly dividends and the store did provide a wide range of
services apart from shopping, including such things as dress-
making and dentistry, and an optician called each month. The
store ran an education department which offered entertainment
and reading facilities in the store hall and it organised annual
excursions. In 1908, for example, there were excursions to
Edinburgh at 5s. 6d. for adults and 2s. 9d. for children. They
occasionally ran competitions, too. In 1912 they offered prizes
of £5, £4 and £2 in a pipe-colouring competition. In 1917 they
opened a penny savings bank for members at 4 per cent interest,
and one of the major services they provided was the extension
of credit. They ran a delivery service around the neighbouring

villages and up and down the pit rows making shopping a very convenient business for the wives of Throckley.

As a business enterprise the store was very successful as the figures in Table 4.2 indicate. In the sixteen years leading up to 1920 membership doubled, sales increased eightfold, profit increased fourfold. By 1920, in addition to its own assets, the store held £46,785 worth of investment capital in other companies and was making annual subscriptions to a wide range of organisations such as the National Life Boat Institution, the National Council for Civil Liberties and the Central Labour College.

Table 4.2 Business records, Throckley Co-operative Society, 1904-20

	Membership	Average quarterly sales £	Quarterly profit £	Accounts owing £	Accounts owing per member £	Dividend in the pound £
1904	1,207	13,565	2,624	575	0.47	3/4
1908	1,403	20,739	3,820	931	0.66	3/7
1912	1,657	39,594	6,440	1,717	0.96	3/4
1916	2,071	33,504	9,726	2,357	1.13	3/-
1920	2,433	110,947	11,091	1,639	0.67	2/-

Source: Balance sheets of Throckley District Co-operative Society Ltd, North Eastern Co-operative Society, Gateshead.

Membership of the store executive committee was a mark of some honour and, particularly in the period after the war, a matter of local political concern. Some of the Labour men - George Curwen and Billy King, for instance - were on the co-operative committee in 1905 although the Liberals led by John Eggie were still the dominant group. By 1913 the position was reversed.

Work in the local store was eagerly sought by many believing it to be secure and respectable. Indeed, my grandmother's niece and later her own daughter, Eva, worked for a while 'in the stores' and so did Francy, aunt Maggie's daughter. My grandfather's view of the store was mildly cynical. He used to say, 'One son in store is worth ten down the pit', implying that the scope for fiddling was great, but his views were not sufficiently strong to incline him to leave.

The significance and meaning of the store goes well beyond the service it provided, however. Co-operation for the activists was much more than simply a way of shopping; it implied a whole social philosophy. The following is a report in the local press of the Lib-Lab MP for the area, Charles Fenwick, addressing Newburn store, the sister store to Throckley, at its annual meeting in 1905:

They were making handsome - too handsome he thought -
returns to the members each quarter. He was not a believer
in big dividends. (cheers) By having reduced their tariff
they enabled working people to continue their connection
with the Society when times were bad.... He would like to
see the productive side of the movement develop more....
We are a long way from the realisation of State Socialism in
this country, but in the co-operative movement they had a
form of voluntary socialism which interfered with no man's
liberty and did no injustice to any man, and yet tended to
increase the material resources of the working classes, and
thereby to considerably encourage their opportunities for
enjoyment.
    He wanted to see more attention given to the educational
side of the movement, particularly among young people.
Instead of devoting so much time to football, let them give
more attention to education. ('Blaydon Courier', 8 April
1905)

That same year John Wilson, the Durham miners' MP addressed
the annual general meeting of the Throckley store. He spoke
after the acting chairman, school teacher and local Liberal
activist John Eggie, had referred to Throckley as 'one of the
best little societies in the North of England'. 'He thought he was
a poor man who worked, as some co-operators worked, for
dividend.' Co-operation, in his view, had a wider ameliorative
social purpose:

Dividend was a sweet thing but he believed this to be the
mercenary side of co-operation, and not the real side....
There was a higher ideal - a force to remedy the social dis-
abilities under which they were labouring. ('Blaydon Courier'
14 October 1905)

The newspaper notes that, at the end of the meeting, 'songs
were rendered' by Miss Todd of Haydon Bridge.
    Dividend accumulation was the only form of saving available
to people in Throckley. In 1913 average purchases amounted to
£24 per quarter and dividend was paid out at a rate of 3s. 4d.
in the pound. This means that an average customer would
accumulate approximately £4 each quarter, £16 each year, a
considerable sum of money before the First World War. This was
a time when the store advertised men's suits at £2 5s., cycles
for £4 19s. 6d., a grey overblanket for 10s. 6d. and a 'Genuine
Grandfather Clock' for £3.
    My grandparents were members of the store from the time they
arrived in Throckley, and through his pig rearing my grandfather did business with it, selling his pigs to the store butcher.
He was not a co-operative ideologue; the store for him was a
means of saving, a source of credit and a supplier of reasonably
priced goods. It was not, however, just a shop. He took an

interest in its affairs and he supported strongly the idea of distributed profits through dividend and the provision of a wide range of services for the community.

I conclude this chapter with a point of theory. The institutions described - the coal company, the churches, the school, the political parties, the co-operative store - and the various clubs and events which were the warp and weft of Throckley's associational life can not be understood as fixed structures in a metaphorical social landscape. For the people who lived in Throckley these institutions were woven into the daily patterns of their everyday lives. They carried a distinctive significance for precisely that reason. The meanings which were attached to these institutions by the people of the area represent the symbolic boundaries of Throckley as a community (cf. Thorpe, 1970). From this perspective, Throckley was not a community bounded by the dilly line and the fell road; its boundaries were much less evident. They were defined by the shared system of meaning and values which people from Throckley could draw upon to give a coherent account of their social life.

In the next four chapters, through a discussion of my grandparents, I seek to show how that pattern of meaning was maintained in the daily routines of everyday life, at work in the pit and in the home, at play, and in the character of relationships in the family itself. But what the account shows, necessarily, is that the mosaic was a changing one; Throckley was a constructed community, but unlike a building it was never finished. It was always, as it were, in the process of construction and, to press the metaphor, under the guidance of different architects at different points in time.

5 PIT WORK

Throckley colliery was started in 1869. It stands where the old flood plain of the Tyne gives way to the rising valley at the base of the Well Field. The colliery had two shafts, the Isabella, named, as I have mentioned, after the daughter of one of the coal owners, and the Derwentwater, named, perhaps, after the Earl of Derwentwater, a former landowner in the area.

The expansion of the colliery in the last decades of the nineteenth century was what lay behind the growth of Throckley itself, and it is to a great extent the nature of the work which went on there which explains the character of the community which the pit spawned. Here I am in full agreement with George Evans who perceptively noted (1976, p. 152),

> to understand the basic structure of the social relations in a working community we have, first and foremost, to study the work itself in some detail; in other words we have to know the material culture at least moderately well.... For a man's attitude to his fellows grows, at least in part, out of the terms and conditions under which he works.

In what follows I seek to describe the pit, the work done in it and the kind of men who worked there revealing, I hope, something of the reality of pit work and, by that token, a major part of the experience and the class position of my grandfather and men like him. What the account shows is that work in the pit opened up questions of politics and economics which went far beyond the immediate setting of Throckley. The union was a critical means of political education, opening the eyes of miners to the larger world around them, forging from their immediate experience of work a consciousness of class and social affairs which nourished the Labour Party and which ultimately eroded the dominant position of the coal owners.

Class is a relationship between men. According to circumstance and opportunity it is a relationship marked at one extreme by harmony and co-operation and at the other by conflict, acrimony and struggle. I show in this chapter that while conflicts born in the act of winning coal were pursued at different levels in the social structure, e.g. in union negotiations with employers or in campaigns to influence government policy, the politics of such conflicts cannot be understood apart from the character of graft underground.

Pit work shapes the men who have to do it, casting their

characters in a special mould. It builds a self-respect and social status around such values as toughness, endurance and underground skill. 'It is clear', write Dennis, Henriques and Slaughter (1956, p. 74), 'that the work a miner does and the wage he receives both express concretely his status as a man and as a member of his profession.' The work builds up, too, a basic attitude of helping others; it devalues competitiveness for the conditions are too dangerous. It fashions a distinctive pattern of social relationships both in families and in communities which must be understood if the more obvious political and industrial attitudes of miners are to be explained.

Pit work must not be seen, however, in isolation from the organisation of the companies which employed the men, the markets for which they produced coal or the specific geological problems faced in particular pits. Finally, it cannot be divorced from the character of economic life in the areas where the pits are found. Coal mining, as I explained, is a variable process and pit work differs from one pit to the next. The point that Royden Harrison (1978, p. 14) made with respect to the study of mining trade unionism, i.e. 'that we need more historical micro-comparative studies of coal mining communities if we are to return again, with profit, to histories of coal mining trades unionism', applies equally forcefully to pit work itself.

THROCKLEY PITMEN

Like the Heddon Margaret, Throckley was essentially a local pit recruiting the bulk of its labour from the immediate vicinity.

Table 5.1 Throckley miners, 1871, by birthplace

Place of birth	No.	% of total
Throckley	0	0.0
Villages nearby	55	34.1
Villages or towns in Northumberland	51	31.6
Elsewhere in England	37	22.9
Scotland	9	5.5
Ireland	9	5.5

Source: 1871 census

The birth places of those describing themselves as miners are shown in Table 5.1. The data are taken from the census enumeration sheets of 1871 for the Throckley district. Only about

one-third of the workforce came from outside Northumberland
and a very small proportion of those came from Scotland or
Ireland. As I have shown in chapter 4, they were housed by
the coal company in accommodation thought at the time to be of a
high standard.

The coal company had a clear policy of recruiting what they
thought of as 'good workmen'. They were interested in respectable family men and particularly in those with Methodist convic-
tions and abstemious habits. A report of 1897 describes something of the character of the men found at Throckley, commenting especially on the drink question:

> It is estimated that the number of inhabitants now reaches
> close upon 2,000 and for a pit village it is claimed that in
> the orderliness and prosperity of its people it is second to
> no other in Northumberland and Durham. More than half of
> the miners are total abstainers, for the prohibition of drink,
> dogs, and pigeons keeps away those that are inclined there-
> to, and the result of the repulsion of these free-livers is a
> sort of artificial selection of steady workmen, who have in
> the course of years formed themselves into an industrious,
> peaceable and thriving community, as is evidenced on every
> hand. ('Newcastle Daily Chronicle', quoted by Hayler, 1897,
> p. 25)

The report goes on to note that both chapels in the village were
well attended and that the number of men attending morning
service in the chapels was exceptional:

> The miners are equally good in turning up for work in an
> efficient condition on Monday mornings - guiltless of 'after-
> damp' from any Saturday and Sunday potations - and it is
> said that Throckley Colliery leads all the collieries of the
> county in this respect. (ibid.)

And with reference to the fact that the Board of Guardians for
the district were only required to meet the school fees of three
out of five hundred pupils, the report concluded: 'Pinching
poverty is almost unknown in this healthy and well-conducted
village.' The enthusiasm of this account must be tempered by
the acknowledged facts of a high infant mortality rate, insani-
tary drainage and cramped living conditions described earlier.
But it does seem that, given the standards of the time, Throckley was a good mining village, recruiting a respectable class
of workmen who were attracted, we might surmise, amongst
other things by steady work and comparatively good social
facilities.

The pit produced for an inland, largely local market, and up
to 1914 expanded steadily. It was free of the intermittent periods
of expansion and contraction which affected those collieries
elsewhere in Northumberland which were much more dependent

on the export trade. This expansion is seen in Table 5.2. The figures show that for every hewer there were at least two other men (or boys) employed. Other classes of workmen included putters, pit officials, engineers, blacksmiths, stable hands and transport men as well as those who worked at the small coke ovens attached to the pit.

Table 5.2 Numbers employed by Throckley coal company in Throckley colliery, 1898-1914

	Men employed	Hewers employed (in March of each year)
1898	724	–
1899	798	–
1900	850	321
1901	852	332
1902	835	328
1903	833	319
1904	873	343
1905	944	343
1906	874	370
1907	917	–
1908	906	–
1909	912	–
1910	874	–
1911	896	–
1912	901	–
1913	921	–
1914	772	–

Source: Northumberland Coal Owners' Association, Statistical Information, NRO NC8/C/1.
Note: The statistical records are incomplete and no colliery records exist.

Steady employment means, above all, steady wages and a sense of security and an ability to plan ahead. But it also implied that the coal company was a good one to work for, a factor contributing directly to that subtle balance of reciprocity between the owners and the men in which in return for a steady job the men gave a certain loyalty to the company. Other factors were, of course, involved, particularly the way in which the men were actually treated, but the steadiness of employment in Throckley was one of the pillars of a rather harmonious structure of industrial relations.

The unwillingness of Throckley men to strike and their general view that this was a good company to work for might also be explained, partly at least, by the fact that labour turnover in the pit was comparatively low; Throckley men, in contrast to

those in pits further north, had, because of good facilities in
the village, a vested interest in stability. Sam Scott, the present
general secretary of the Northumberland Miners, suggested this
to me while making a comparison with pits in the Ashington area.
Miners in that area were, he said, 'just like gipsies, moving
from one pit to the next'. 'Even the chickens and geese', he
said, 'used to lie on their backs every three months with their
feet in the air ready to be tied up, put in a poke, and moved
on to the next job.' In addition, Throckley was a comparatively
small colliery so that pitmen could much more easily get to know
one another well. There were few strangers in Throckley.

WORK UNDERGROUND

The pit was well placed geologically to work most of the main
seams in the western district of the coalfield. It produced coals
from the Harvey seam (known to the pitmen of Throckley as
the Engine seam and always pronounced with a long stress on
the 'i'), the Hodge seam, the Tilley, the top and bottom Busty,
the three-quarter seam and the Brockwell. This range of seams
gave them access to several different markets. They produced
mainly household coals which were sent to London, some coking
coal, and steam coal which they sold to Spencer's steelworks
and to the railway company. Some gas coal was produced and
sold to the Elswick Gas Works of which Sir William Stephenson
was the chairman.
    The pit presented no special geological problems. It was in an
area peppered with old and even ancient coal workings which
the men would occasionally break into (hole into). It was a wet
pit lying low in the valley and 1,200 gallons of water an hour
had to be pumped out, draining away through the pit pond down
the 'burn' to the Tyne. To the west of the pit the coal measures
outcropped; to the east, below Newburn, they gave way to deep
sand and gravel; to the north they were displaced by the
igneous intrusion of the Whyn Dyke, a massive fault with a
70-foot displacement. The seams themselves varied in thickness
from 18 inches to 3 feet, and although the pit was wet working
conditions were not considered to be too bad.
    This geological information was given to me by one of the
former engineers of the colliery, Bob Reay of The Leazes; but
the same facts concerning the dampness of the pit were interpreted for me in quite a different light by Mrs Thompson, the
daughter of the union compensation secretary, Jack Ritson. She
remembers as a girl watching miners returning from work in the
winter with their wet clothes freezing as they walked, and
throughout the year pit clothes having to be dried each day
around the fireplace.
    The system of mining coal in the Isabella right up to the 1930s
was the traditional pillar and bord system with ponies being
used for haulage. This was a system of working in which miners

achieved a great deal of functional independence and in which small groups of mates or 'marras' could work together, determining within wide limits the level of their own output and, therefore, their own earnings. A description of this system has been given by Eric Wade, a former mining engineer turned sociologist. I quote fully from Wade since this method of working required of the miners a special range of skills, and as the payment system underground was based on the different kinds of task this system of coal winning required, it is important to be aware precisely what the system was.

In the Northumberland and Durham coalfield there were two basic methods of working that required the use of putters. First, there was the Bord and Pillar system pioneered in the Tyneside coal basin and secondly the Longwall Gateway System. The Bord and Pillar system was at the first stage of working a partial extraction system. From the main roadway entries were made into the coal seam to form a flat or district. These entries were driven at such a dimension as to allow the passage of a tub. These entries were known as tramways. There were two kinds of tramway (a) the Bords, and (b) the Headways. The Bords were driven at a greater width than the Headways. Coal has a basic cleavage plane known as a cleat. Hewers on working the coal found it easier to work in a direction at right-angles to the cleat than in a direction parallel to the cleat. Consequently for a given period of time more coal could be won in a working place in a Bordways direction than in a Headways direction. The Bords were driven, usually at least one yard wider than a tramway travelling in the Headways direction. At the first stage of working an area of coal would be locked out, leaving pillars, the dimensions of which were determined by the depth of the coal-seam from the surface, and the strength of the surrounding rock and the strength of the coal. As a rule of thumb between 40% to 50% of the coal was obtained in the first stage. This was known as working the 'whole.' When a predetermined boundary was reached, the pillars were removed retreating towards the original entry point of the flat. This was known as working the 'brokens.' (Wade, 1978, pp. 24-6)

As a method of mining the pillar and bord system gave rise to little dust and it was relatively quiet. But it was very wasteful of coal since pillars of coal had to be left for safety reasons, and it did involve taking out a lot of the roof to expose the coal. In the absence of mechanical conveyors and continuous coal cutters it was, however, the most technically efficient system of winning coal.

Reliant on physical labour, the system shaped the sequence of tasks (or the task structure) the miners were required to perform. A brief description of these tasks and skills required for their performance indicates the complexity of the miner's skill.

Once the 'district' or 'flats' had been mapped out the miners' task was to drive the roadways, the headways and the bords. This involved taking up bottom stone and taking down roof space. Once the space was cleared supports had to be fitted, the wood needed to be precisely cut and shaped, and tramways laid down.

Working the coal itself involved drilling, the controlled use of explosives, heavy and inconvenient work with a pick, under-cutting the seam of coal (the 'jud') and either shooting it down or hacking it away. This process of undercutting was known as 'kurving'. Once cut, the coals had to be 'filled' using a large pan-shaped shovel (always referred to in the Throckley area as the 'pan shull'). The filled tubs of coal were then pushed to the main tramways by the putter and transported to the pit shaft. After haulage the tubs were weighed by the master's checkweighman and the miners' checkweighman, a man whose wages were made up by the men themselves.

Varying geological conditions - e.g. faults or hard coals, crumbling walls or hard stone floors - could affect output seriously. Payment for work depended upon how much coal was produced. Therefore, to make the distribution of working places equitable they were 'cavelled' for each quarter, i.e. there was a random allocation of work places among hewers and putters. And to acknowledge that a great deal of preparatory work was necessary before coals could be won, the agreements struck between workmen and employers contained separate prices for different kinds of job.

Pit work is highly skilled with a finely graded career struc-ture from trapper to putter and then, at the age of 21, to hewer. There was, of course, no formal training; essential skills were picked up, often being passed on by fathers to their sons. The first thing a miner had to learn was the geography of the pit. Miners were never too sure exactly where they were under-ground in relation to the surface. Time and distance are dis-torted by darkness and it takes time to know the way under-ground although it is essential to get to know it if only for escape. The men had to get to know about the flow of air, the problems of gas building up and the dangers of naked flame. This was a matter of knowing the law. Under the Coal Mines Regulations Act of 1872 many working practices were illegal. Boys under 18 were not permitted to be in charge of dangerous machinery, pit cages had to be properly covered, travelling roads had to be provided with man holes, pit props had to be fitted properly, water levels had to be regularly inspected and so on. These rules were for the general safety of the pit and had been hard won by miners themselves, but their violation resulted in court-enforced penalties for both employers and men. These regulations were enforced by the Mines Inspectorate and they had to be learned and understood.

Pit work is a strenuous art requiring not so much that a man should be physically well built but that he should be sinewy and capable of sustained effort over long periods. My grandfather was uncharacteristically tall. He stood over six feet but he was always lithe and strong and had huge working hands. Pitmen have to be agile, too. To get to a coal face a man has to walk slightly stooped, often for up to a mile; he has to dodge rock protrusions and perform a number of actions - swinging a pick, shovelling, ramming props into place - in very cramped conditions.

Finally, pitmen need to understand the precise roles of others, to know what to expect from other men, to learn that men can be trusted and to acquire, as a matter of almost instinctive response, an ability and willingness to help others below ground. Without that no pit can function properly.

It is difficult for non-miners to appreciate that such work, despite its difficulties, can bring satisfactions. My grandfather, like many others, was proud of his skills as a hewer. He enjoyed the company of other men. He liked the conversation - 'the crack', as he called it. He took great care in his work and enjoyed seeing jobs done properly, with precision. And he valued his autonomy. Eric Trist and his colleagues have referred to the miner under the pillar and bord system as a 'complete miner' performing a 'composite work role' (Trist et al., 1963, p. 33). It is this composite work role which 'has established the tradition and reality of faceworker autonomy' (ibid.). Since the same degree of autonomy was not available to putters in the pit, they being much more strictly controlled by the underground officials, it is understandable that the coal hewer, having experienced close and unwelcome supervision as a young man, would value his freedom and the underground status that freedom brought.

ACCIDENTS AND RISKS

Miners also need to acquire that uncanny knack (in other professions it would be called an art) to anticipate dangers and balance their efforts at winning coal against the contingencies of the immediate situation. My uncle Jim emphasised this to me in this way:

> No, you didn't have fears; you get used to it. You are nervous at times. If it's dangerous you protect yourself because you know you've got to, and you keep her well timbered. Otherwise when it's good you seem to neglect it, you must go on you know, she's all right. And that's where people get hurt. You get a fall when you are not expecting one. But when it's bad and you know it's going to fall you timber it up to stop it, where when it's good you never think about it. That's where people gets lamed.

The subtlety of such interpretation is lost on miners themselves;
they live in a taken-for-granted world but they are practising
an art, nevertheless. In my grandfather's day the risk of fatal
accident may not have been high but the risk of injury which
could interrupt earning was high. The figures in Table 5.3
indicate what the risks really were.

Table 5.3   Accidents in the Northumberland coalfield,
            1899-1933

	No. of fatal accidents	Proportion per 1,000 miners	No. of non-fatal accidents	Proportion per 1,000 miners
1899	30	0.79	3,174	84.2
1900	52	1.30	3,107	77.4
1905	47	1.04	4,961	109.9
1910	42	0.75	8,107	144.2
1915	47	1.15	6,305	154.4
1920	36	0.59	5,168	84.4
1925	82	1.51	8,635	159.1
1930	49	1.06	8,525	184.9
1933	34	1.13	5,840	194.3

Source: Northumberland Coal Owners' Association, Mutual
          Protection Association, 36th Annual Report, 1933,
          NRO.

The more effectively the art is performed the lower are the
inherent risks of working underground. But there are risks,
nevertheless, and they have to be faced. Harold Heslop in
'The Earth Beneath' noted a feature of the character of many
miners when he wrote: 'Fearless contempt is the basic attitude
of all those who chisel from the earth any of her treasures'
(Heslop, 1946, p. 27). Without some belief in fate and a resigned
acceptance of danger it is not possible to work underground;
fears need to be repressed and the miner needs to trust his own
abilities. Small accidents were regular occurrences, bruised
back-bones, cracked shins, cuts, scrapes - all of them leaving
the characteristic blue scars which are to be found on the skin
of anyone who has worked underground. Fatal accidents were
less frequent but happened often enough to be a worry.
My grandfather himself had an almost fatal accident in 1913
which laid him up for a while and left him with several permanent
scars on his hands, arms and back. Jim described the accident
to me in this way:

He was hewing and a jud fell on top of him. That's how he
had all the blue marks on him. He would have been dead.
There was an old fellow called Tommy Urwin - he went by
the name of Barmy. If it hadn't been for him he would have

been a dead man. But me father can remember seeing the
light coming around the turn to come into his place then the
jud came over. Barmy heard the jud and he run and he
scratched down among the coals. He got him out and saved
his life. He was cut all over. But he was off work a canny
bit.

My grandfather said no more about it than he was lucky; he was
grateful to Barmy Urwin but he never felt obliged to him in any
way as a result of it. He himself would have done the same had
the occasion arisen.

My uncle Bill, a schoolboy at the time, heard of his father's
accident before my grandfather had reached home on the back
of the colliery cart; the news had spread up the coach road and
reached the schoolyard. This underlines starkly that what went
on in the pit was very close to everyone, a point brought out
even more vividly in the case of fatal accidents.

If there had been a fatal accident the pit buzzer blew so that
everyone knew about it immediately. What nobody could know
was who had been killed; that news came later. For a while,
therefore, that buzzer sounded doom for everybody and symbolised in a fearful way the presence of the pit and its dangers.
When there was a fatal accident it was customary for the pit to
cease production and there was always a collection among the
men to help the widow.

The effect of death in the pit was a profound one; it forced
everyone to consider his own personal position, thereby under-
writing the commonly heard phrase, 'There but for the grace
of God go I', and it symbolised starkly the collective interests
of the whole community and its dependence on the pit. More
profoundly still, by interrupting the busy flow of routines, a
death in the pit gave people a chance to reflect and to compare,
to distance themselves from their everyday life, if only for a
while, and to see it in a new light, connecting it with the past
and relating it to the collective history of miners as a class.
Harold Heslop notes of the Hartley disaster, for instance, that
it 'enriched their conception of death' (1946, p. 161). The
disaster helped the miners to identify with the role of 'the
maligned creature of injustice'. 'Hartley', he writes, 'urged an
inward pity which evoked a newer and more palpable resolution.... A conception of justice springs from lowly imaginings'
(ibid.).

In my grandfather's lifetime there had been terrible disasters
at Seaham and the 'death pit' at Stanley. Talking about them
gave him the opportunity to emphasise his view that it would be
a 'good job if nobody ever had to work doon a pit'.

This view was reinforced for my grandfather when, in 1906,
William Breckons, the son of his sister Alvina, was killed at
Heddon. The funeral performed an important function, too. It
brought families together and strengthened communal ties in
the village. My mother told me:

We took death seriously then, much more seriously than now.
There were big funerals led by two 'white hoods' - I was one
once - and the procession walked to the church. People went
into mourning for a long time ... for often up to three months,
even the children wore black armbands.

If the death occurred in the pit the cortège was often led by the
pit brass band. Funerals are occasions for reflection on the ulti-
mate significance of our being on this earth and an opportunity
to consider how important particular relationships have been and
to re-think basic questions of religion.

My uncle Bill told me that he and my grandfather went to a
funeral in the pit village across the river, Clara Vale. The day
was wet and the ground very muddy. The lowering of the coffin
into a sludgy grave upset Bill and afterwards, angry at the
meanness of it all, he said to my grandfather that it would have
been just as appropriate to leave the body in the pit. My grandfather reacted badly to this, saying, 'If you did that we'd be
no better than animals; you've got to have a proper burial.' Bill
says he felt humbled by this reaction but it did not change his
sense of disgust.

The records of the Northumberland Coal Owners' Mutual
Protection Association coldly records cases which would almost
certainly have been discussed by my grandparents and about
which they would feel 'There but for the grace of God go I'.
For example:

Case No. 387 Throckley

> Matthew Hope, 43, deputy. Died on May 21st 1907
> from acute pleurisy.
>
> The doctor who attended Hope certifies that death
> was the result of injuries received on April 8th 1907
> when Hope was crushed by a fall of stone.
>
> Deceased leaves a daughter aged 10 who is the only
> dependent.
>
> Three years earnings   £274 7s 3d
> House and coals   £ 39 0s 0d
>
> Deceased received weekly sum of £4 in the form of
> weekly payments of compensation prior to death.
>
> Claim made.
> Full liability admitted £296

Or, a case much closer to my grandfather's condition:

Case No. 463 Throckley

Matthew Nixon, 31, hewer. Killed on September 25th
1908, by a fall of stone.

Deceased leaves a widow and one daughter aged 5,
wholly dependent

Three years earnings   £277   0s 8d
House and coals   £ 33 16s 8d

Claim made by widow. Full liability admitted £300

The widow in each case would as a matter of course have to
forfeit the house.

In the case of minor accidents there were compensation payments, but the real problem here was the inadequacy of medical
treatment. Throckley colliery did not acquire an ambulance until
1907; before that stretcher cases were taken along the West Road
to Newcastle General Hospital on a flat cart pulled by a horse.
If there was an accident requiring such transport the horse had
first to be untethered and brought down from the Well Field to
the pit. In 1907 the coal company donated £150 towards a small
local hospital (which itself had been set up on the initiative of a
group of local doctors and which was supported, too, by miners'
union funds) but urgently needed treatment could not be guaranteed.

Miners needed to be conscious of their health; it related directly
to their output and therefore their wages. They placed a value
on being big and strong and fit for precisely this reason. As I
shall show, these same values were held by their wives.

RELATIONS UNDERGROUND

Relationships underground were of two sorts, those concerned
with other miners and those involving officials. Among the men
themselves in Throckley there was a general friendliness and
basic solidarity; the pit was thought to be a happy one and most
men were known to one another for, as I have explained, there
was little labour turnover there. There was, none the less, a
subtle hierarchy among the men and many of the social differences
of the surface, such as church affiliation or views on drink,
particularly where men came from, reappeared underground in
an attenuated form but had a powerful influence on who became
workmates. There were some men my grandfather would not work
with because he thought them too lazy or unreliable.

Before his sons started work my grandfather worked at Throckley pit with three 'marras', Mr Guthrie from The Leazes, Watty
Barnes from Blucher and Mr Watson from the High Rows. Like
most other groups of men they worked together over many years,

building up a relationship of trust and mutual respect. They had
to rely on one another for their output and safety. Their earnings depended upon each one pulling his weight, turning in for
work, making the place safe and getting out the coals. The men
were paid as a group and the foreshift pair would get the wage
every fortnight and share it out among the others. The pay-out
was on an equal shares basis unless there were adjustments to
be made for absences. These men were good friends of my grandfather; Mr Guthrie had an allotment next to his so they met
regularly outside the pit. But their friendship was a functional
one for the pit and the coal company. The relationships within
the group itself acted as a great force for discipline at work.
Any slackening, any unwarranted absence, any slipshod work
would bring the appropriate response from the others, either a
mild chivvying or a severe rebuke. It was knowing that, though,
which kept them all together working hard for their own collective good, and they greatly valued being able to work with
others who worked well. There was a subtle hierarchy here based
on ability to work which reinforced among the men themselves
those qualities of effort and practical perseverance which the
company valued in their employees. As a pitman my grandfather
valued men who would 'pull their weight'. Doing what was
expected of him at work was for him a positive value and he
expected others to do the same. This was an attitude which carried over from the pit into other areas of his life: his avoidance
of charity, his dislike of 'scivers' and his determination to be
independent. He placed great faith in what he took to be the
fact that if he did his job well he could be free of carping control underground, that his own hard work was ultimately his
best defence.

But there is one relationship which, above all others, deserves
special note in the pit, that between father and son. There was
no formal apprenticeship in mining although clearly there was
much to learn, and this learning was frequently passed on from
father to son. Fathers had an additional role: that of controlling
their sons in the pit and of making sure the basic discipline of
working - getting up on time, being ready for work, obeying
the rules and giving their effort - was maintained. In 1902 the
Executive Committee of the Northumberland Miners' Association
discussed a case from Throckley of a miner being dismissed
because of his son's behaviour. The minute reads: 'We are
unable to entertain the case of the member of this colliery who
has been discharged in consequence of a misdemeanour committed
in the mine by his son' (NMA, Executive Minutes, 1902, NRO
759/68). What seems clear is that the union tacitly agreed with
the right of management to act in this way. In doing so the
union was acquiescing in a form of management control which
exploited kinship to achieve its end of regulating men at work.
The way this worked out in my grandfather's case comes out
clearly in my uncle Bill's comment on smoking down the pit, an
offence punishable by imprisonment.

You weren't allowed to smoke in lamp flats ... but there'd be
a lamp flat with a great strong current of air. We as miners
knew that it was impossible for gas to lie and we'd maybe
smuggled some dumpers in and smoked them in the airway.
If my father heard about it! I remember once, me father an
me. He said, 'You've been smoking.' I said, 'No.' Whey he
could smell it! He didn't half give me a bloody lesson. He was
strict, lad.

And uncle Jim reinforced this point, saying:

He played war with ye. Aa've seen him come into a place
after we've went oot like and he's followed us in. If it hesn't
been timbered properly he's played war with us coming hyem
at night aboot the way we've left the place, left dangerous.
He'd play war.

When I raised the question of attendance and absenteeism with
them they answered, almost in unison, 'Absenteeism! There was
none!' And Bill went on to relate the following tale which gives a
vivid account of my grandfather's attitude to work:

I was once on the back of a motor bike and we hit a drove of
horses. I knocked me thumb oot. He was sitting with his pit
clothes on when I come in. 'You're late.' I'll never forget
this. He was sitting with that bloody pipe in. I said 'Aye, a
cannot gan, father.' 'Bloody pleasure, loss of work.' He
started at aboot twelve o'clock at neet. And here's me thumb
oot. I went to the doctors next morning and he asked how
long I had had that and how did I suffer it. I said 'Ye want
to gan and ask me father. He wanted me to gan to work.'
Work was first with father, pleasure second. If there was
only time for work there was no time for pleasure. It was as
simple as that.

Underground working was governed by a very complex system
of payments and appeals and is well illustrated in the following
agreement, made in joint committee, between the coal company
and the workmen at Throckley in 1908. I quote it in full because
it illustrates both the terms of the employment contract and the
kind of issues about which the miner had to keep his wits vis-à-
vis the underground management.

All seams, except Hodge seam, to be worked by curving juds
and nicking and to be wedged and shot down. No jud will be
allowed to be shot down unless it has been nicked and curved
to a depth of not less than 30 inches. Scalloping will not be
permitted in any of the seams except the Hodge seam. Best
coal to be filled by itself and splints and top coal may be
filled together.

Nicking, curving, scalloping, wedging and shooting are the basic processes of cutting the coal from the seam and were all carefully regulated for reasons of safety and coal quality.

Hours Foreshift hewers will go down pit at 4.0 a.m. Back-shift hewers at 10.0 a.m. All hewers to change at face. Lads to go down pit at 6.30 a.m. and ride at 4.30 p.m. Back-shift men to ride after lads. Pit to commence to draw coals immediately lads are down and cease at 4.30 p.m. On Baff Saturdays, pit will hang on at 4.0 a.m. and cease at 2.0 p.m.

Throckley men worked a six-hour shift underground and the lads an eight-hour shift.

Rent Hewers in rented houses and working ten days per fortnight will be paid 2s per week house rent and hewers working less than ten days in a fortnight will be paid at rate 4d per shift.
Deductions Coal leading 6d per fortnight. Water rate 4d per fortnight. Broken lamp glasses 6d each. Broken window panes cost of same. Repairs to drilling machines, cost of same. Lost tokens ½d each.

There were, in addition, deductions for the union and the cottage hospital; the men had to pay for pick sharpening by the pit blacksmith and they had to buy their own powder from the company. Coal produced was paid for at given rates per ton and each particular task, e.g. winning headways, was paid for separately, the manager measuring the yardage won by each man and agreeing a price there and then to be paid for the job. Putters were paid partly by the distance they had to push their tubs. Some faces were a long way from the main roadways and this distance had to be compensated for both by the cavelling rules and the price arrangements.

The point about these complex pricing systems is that they invited conflict underground with constant haggling over distances and whether particular places were too difficult to work profitably. In Northumberland those differences which could not be solved at pit level were referred to the joint committee of owners and men. It had the right to appoint 'referees' who would go to the pit and investigate the dispute, having in many instances the power to settle it. Most of the time, however, the problems were handled at pit level.

It is this bargaining which above all else sharpened the consciousness of miners about the broader context of their work and which forced them to acquire, in addition to their technical skills, an economic wit and negotiating expertise which was vital to their earning power. George Ewart Evans has noted the consequences of such arrangements:

Negotiation, skill in bargaining, was ... vital to them in order
to avoid the sharp edge of exploitation. They had a long and
hard training in this, and in countering the numerous kinds
of manipulation that were used in an attempt to sell them an
agreement that was not basically in their interest. This is the
reason why the miners have so often become a stumbling
block to an unsympathetic government. (Evans, 1976, p. 157)

The whole system led to subterfuge, collusion, double standards,
manipulation and conflict, each intensifying or decreasing according to the market price of coal and the capitalist's sense of his
profit. For the working miner, however, knowledge of the agreements and an acute awareness of his entitlements were essential
parts of his skill and of his sense of his whole social position.
His education in political economy was hard won in the literally
black business of getting coals.

Relationships with management were of two sorts, those with
the coal owners themselves and those with pit officials. As
already explained, the Throckley coal company, like many others,
and certainly like the large landowners of England, delegated
the responsibility of running the colliery itself to their agents.
This policy was for them functional; it meant that work-generated
conflicts were always absorbed in the first instance by the underground management, the viewer, leaving the coal owners to bask
in the glory of their paternalism. And the fact that the owners
themselves did not own the land or the coal, only buying through
royalty payments the right to mine it, meant in addition that they
could always project with some credibility the idea of a common-
ality of interests between themselves and their workmen as
against the royalty owners. Playing very much the role of the
third party in industrial relations at Throckley, the Stephensons
also managed to cast themselves in the role of peacemaker.
Several company employees have stressed for me the fact that
men would often go over the heads of the manager of the pit to
take their grievances to the major, finding him to be invariably
fair and willing to listen.

The relationship between owner and workman contrasts sharply
with that between workman and manager or underground official.
Miners in Throckley did not accord underground officials the
same respect or tolerance. The underground staff were invariably the butt of the miner's discontent and filled this role for
two reasons, one concerned with the social relations of employment, the other with the nature of a miner's work. The agreements struck between the Northumberland Miners' Association
and the Northumberland Coal Owners' Association - agreements
which reflected the general state of the coal industry as a whole
and the relative power of the two groups - actually made some
form of conflict underground inevitable. Conflict was built into
the way in which different aspects of the miner's job was priced.
But the work itself, under the pillar and bord system, produced
in the men a powerful resentment of supervision. As it was

highly skilled work and paid according to output and, above all,
carried out by independent groups of marras, the men required
'neither instruction nor co-ordination' (Goldthorpe, 1959, p. 215;
see also Goodrich, 1975). Goldthorpe emphasises: 'Not surpris-
ingly, therefore, the colliers resented close supervision very
strongly; there was a traditional hatred of being "stood over".'

This dislike of supervision was clearly reinforced by the
capricious way in which authority underground could be exercised. Miners could be sacked for using bad language to an
official or for breaking many of the safety rules which the
officials were required to enforce. It was in the interests of
miners, therefore, to so act underground that no charge could
be brought against them. If they worked within the rules they
could keep the underground officials at a safe distance and avoid
any personal contact of a demeaning kind.

Something of the hostility which did exist towards some of the
underground officials is indicated in two accounts of incidents
underground which involved ponies. Old Tom Stobbart, speaking of Matt Foster, the Throckley overman, related a tale in
which a miner was involved in an accident with a 'tickley-back'
(flighty or excitable) pony (pronounced 'powny' in Throckley)
where the bolting animal had derailed its tub and trapped the
driver in a tunnel. Foster's first question on investigating the
scene was 'is the powny allreet?' Tom Stobbart told me this to
emphasise that miners were, in his view, often treated worse
than animals. My uncle Bill told me a similar story involving Matt
Cheesman, a man he detested, shortly after he became manager
at Throckley Maria pit where Bill was working:

He was a horror, a bloody stinker. I'll give you a tale of
Matt Cheesman. This was when he came up to the Maria where
we were working. Well, there was myself, Joe Wade and Togo
Thompson putting at the Maria pit. I was first on and where I
was some were going left, some first right, some straight down
you know. Well, I was going straight down. It was heavy and
we could only pull one tub out. And I had a great big horse
called Nipper. I came out onto the West Brockwell flat.

If we were going to have our baits we would put choppy in
the boxes for the ponies. First of all we would stop and give
them a drink and then choppy. Well, I went to the choppy bag
and had to shake it out into the box; hardly any in and I just
put it down to Nipper. Joe Wade followed me out with his little
pony called Mick. Joe said, 'Any choppy, Billy', I said, 'No,
there's none, Joe', so Togo Thompson followed out just the
same and we sat down at the Deputy's kist. Matt Cheesman
came in. I said, 'Oh here it comes.' I heard him going around
the boxes. Well, mine had choppy in. There was a little in
mine. So he came down and he says, 'Who's puttin' wi' Mick?'
Joe Wade says, 'Me.' He just picked Joe Wade's bait up and
dropped it into Mick's choppy box.

I was always a hot-headed bugger and didn't care if I got

the sack or nowt. And, as I say, I was a bloody big lad.
There was a water trough standing at the side and Aa says,
'Matt Cheesman, if ye had done that to me I would have
ducked you in that bloody water trough.' 'What?' 'I would
have pushed you in that bloody water trough. That lad's
walked from Violet Street, Benwell, this morning, to this
bloody rotten hole and you do that for a bloody pony. I
would have put you in that bloody trough and I would have
drown you.'

Not all Throckley miners saw Cheesman in the same light. Tom
Stobbart told me he thought Cheesman was a fair man, a man who
he used to go 'fleein' pigeons with'. The point here, of course,
is that through education and life-style, being little different
from miners themselves, pit officials were caught up in all the
normal cross-currents of social relationships in the village; some
were liked, others not. Whatever the feelings involved, however,
underground officials were a focal point of conflict simply because
of the employment relationship itself, a point brought out vividly
again by my uncle Bill:

I was twenty-one and a half and coal hewing. My father was
in a heading and I was in a bord. We were taking a bottom
canch up and pulling the tubs through. We hit iron stone.
They had us measured that we were getting so much a yard
for taking it up. We either had to rob the coal company or the
coal company robbed us. I was drilling just top of the iron
stone lifting clay. And, of course, that meant I was taking
less bottom up than I should have been. I bet I was fifteen
yards. The putter had been in and saw the pony scrape the
top a little bit.
   Cheesman came in, *and with my hack,* he pulled all the rails
up. I'm in bye hewing. So he comes, sets himself down. I
turned around and says, 'And you know what you can do?
You can go and lay every one down.' He says, 'You're not
taking your height.' I says, 'Look Matt Cheesman. You are
paying us wrong. You pay us to take out underneath that
iron. You know we can't drill iron stone. It's impossible.
Either you've got to rob me or me rob you. This bord will
only go so far in and it'll be stopped. Which is best?' Him and
me were tooth and nail.
   My father hears us. He comes up. 'What's on here?' 'See
what that bugga's done father? Mind the bugga will lay it down
himself. I don't. Or he'll set the nightshift in because I don't.'
He says, 'I'll sack you.' I says, 'Ye can bloody do that with
pleasure.' Me father got on well with him. 'Now Matt, it's no
good talking to him like that because he'll welcome being
sacked. So don't talk that way to him.'

After my grandfather's intervention the dispute was patched
up and Bill and my grandfather re-laid the rails. But that was

the pit, a daily experience of conflict and acrimony built into
the normal business of winning coal.

THE UNION

It is for the reasons just described that trade unionism for miners
in this period was not something removed from their daily experience and organised by strangers; it was a major part of that
experience. And just as my grandfather insisted on his sons
going to work, he made sure they attended union meetings.
These were held in Throckley co-operative store hall every fortnight or, in emergencies, on the side of the pit heap at any time.
The fact that the coal company was careful in its labour recruitment is partly reflected in the figures for union membership at
Throckley. Since the lodge records prior to 1934 no longer exist
the affairs of the pit have to be reconstructed from the minutes
of the union as a whole. It is clearly an unsatisfactory procedure
but no other is available.

At the turn of the century just over half of the men employed
at Throckley were in the union. But by 1914 the figure was just
on three-quarters with recruitment increasing rapidly during
the troubled period in the few years just before the war. The
figures are as shown in Table 5.4.

Table 5.4   Union membership as percentage of the total labour
force, Throckley Colliery, 1898-1914

	%		%
1898	43.3	1907	58.5
1899	57.1	1908	61.0
1900	55.4	1909	–
1901	55.7	1910	59.0
1902	57.1	1911	59.5
1903	57.2	1912	62.5
1904	53.7	1913	63.1
1905	59.4	1914	74.8
1906	59.4		

Source: From Northumberland Coal Owners' Statistical
Information: Northumberland Miners' Association
(Annual Accounts).

The union as a whole grew rapidly during this period. In 1898
it had 19,000 members and by 1914 the figure had risen to
37,000. Despite this growth there was always a problem, as the
union saw it, of non-members.

The period from 1900 to 1914 was an important one, for the
Northumberland Miners shifted their allegiance from the Lib-Lab
politics associated with Thomas Burt to an explicit support of the

Labour Party; they joined the Miners Federation of Great
Britain and, in 1912, joined in the first national strike of miners
over the minimum wage issue. Industrial struggles in the docks
and on the railways, the formation of the Triple Alliance (of
transport workers, railwaymen and miners to co-ordinate industrial action), the rise of the shop stewards' movement, the
growth of syndicalist ideas and the maturation of the Labour
Party are merely the surface manifestations of a great ground-
swell of protest and discontent throughout the labour movement
which severely challenged the whole fabric of Edwardian society
It is not necessary to agree with Walter Kendall (1973, p. 192)
that 'in retrospect the unprecedented struggles of the years
1911-1914 seem the head of a lance probing the vitals of capitalist
society, demanding recognition of the labour movement's new
found strength and power'; but it is clear that during this
period the veils which concealed the exploitative character of
capitalism were for millions of workers torn away.

For the Northumberland miners this broad transformation in
the position of the working class was crystallised in specific
struggles over the Eight Hours Act and in the minimum wage
dispute of 1912, both opening up the question of whether
Northumberland should join the Miners' Federation of Great
Britain (MFGB) and face, in doing so, the prospect of dismantl-
ing the whole apparatus of district bargaining and, given the
militant stance of the South Wales miners, a long tradition of
conciliatory industrial relations. They were forced to consider
simultaneously their economic position as miners and their political position as unionists.

The early part of this period, up to about 1906, was an
uneventful one for Throckley colliery; union recruitment was
rising steadily with the numbers employed at the pit, and real
wages were rising, too. The only resolution the lodge sent in to
the Executive Committee of the union as a whole was about the
keeping of dogs in colliery houses:

> Seeing that the owners at several collieries in the county are
> forcing workmen either to do away with their dogs or remove
> from the colliery, we suggest that wherever such proceedings
> are taking place that if a majority of the workmen as per rule
> for strike, be in favour of resisting this encroachment of
> their personal liberty, they be supported by the county.
> (NMA, Executive Minutes, 1901, NRO 759/68)

The resolution was, in fact, ruled out of order, but it does
perhaps represent some determination to resist a long-standing
element of the social policy of the Throckley coal company.

Throckley lodge had no representation on the Executive Committee at this time but was represented on the council of the
Northumberland Miners by George Curwen and Bob Hutchinson.
Through them the wider preoccupations of the union were heard
at Throckley and discussed. From reading the minutes of the

union for this period it is evident that the affairs of the mining
industry could never be discussed aside from larger questions
of politics and the state itself, and here, surely, is the reason
why later demands for the nationalisation of the mines can be
seen to have arisen almost naturally out of the workaday experience of pitmen.

In 1901, for example, the union began a protracted battle
against the coal duty which the chancellor of the exchequer had
imposed in his budget. The duty of one shilling per ton export
tax was seen by the Northumberland Executive as something
which would lead to 'the extinction of our industry in the North'.
They objected to it as a direct attack on free trade and called
for a national conference of miners to discuss the option of 'a
complete national stoppage of the coal mines in England, Scotland
and Wales until the tax was withdrawn'. It was eventually withdrawn in 1906.

The campaign for the Eight Hour Day Act, which was passed
in 1909 after years of agitation and against which the coal owners
had long struggled, had finally made it possible for Northumberland to join the MFGB, leading them to accept the case for
nationally negotiated agreements in the mining industry. Throckley had, in fact, voted against joining the MFGB and had voted,
too, against the Eight Hours Act, interpreting this measure, no
doubt, as likely to lead to an increase in the number of hours
they had traditionally worked from six to eight. In the 1906
ballot on this question 131 Throckley men voted for the eight-
hour day and 304 voted against it.

The reason, perhaps, was that the coal owners had tried to
link the passing of the Act to the abolition of the system of free
house and coal, a central element of the employment contract in
Northumberland to which the union was strongly committed. But
what the struggle showed was that against the opposition of
coal owners the state itself could, through industrial action or
the threat of it, be coerced to act on the miners' behalf.

This was heavily confirmed in the coal stoppage of 1912 over
the minimum wage question. Arising out of a dispute in South
Wales - a ten-month-long strike beginning at the Ely pit of the
Cambrian Coal Combine - the 1912 stoppage consolidated the
Miners Federation of Great Britain, radicalised many miners,
especially in South Wales, and finally established the view that
wages were a proper charge to the cost of the industry as a
whole and should be freed from their dependence on the selling
price of coal.

The strike lasted six weeks from 1 March to 10 April and was
very well supported. It was a crucial strike, not so much for the
victory gained in establishing a minimum wage, but for the way
it helped the miners cut through the logic of coal owners that
the living standards of workmen should depend entirely on the
market position of the industry. The coal owners, for instance,
were quite emphatic that minimum wages would wreck the industry. They told a deputation from the miners on 21 October

1911, that:

> the request for a minimum of 30 per cent above the basis
> would have led in past years to the closing of a very con-
> siderable proportion of the collieries in the county, and in
> some years, possibly the whole of them. (Coal Owners'
> Association, Minutes, Book 11, NRO/DC)

And in an open letter to the chief newspapers in the area the
secretary to the Northumberland Coal Owners, Mr R. Guthrie,
warned of the consequences of granting minimum wages, insisting that it 'would, if adopted, be detrimental to the workmen
generally and fatal to the employment of old and weak men'
('Blaydon Courier', January 1912).

This position was countered by many miners with the argument
that, if the industry was more efficiently organised - and this
could only be done through nationalisation - then the costs of
decent wages could be met in higher productivity and greater
efficiency. There is a moving letter to the 'Blaydon Courier' in
the 2 March issue of that year from a Walbottle pitman making
precisely these points.

The strike was successful only in so far as a principle was
conceded; in terms of income it is unlikely that the miners gained
much since the final amount payable, a five shilling per shift
minimum, was below average earnings in many districts anyway
(see Rowe, 1923). It was, however, as the first national strike of
miners, a new kind of strike, and although many miners felt
cheated by the settlement which among other things set up district boards to negotiate wage matters, there was to be no turning back from the idea that the state itself had a key role to play
in running the industry, a further argument reinforcing the
growing demand for nationalisation (see Kirby, 1977). Since
government vacillation was associated with a Liberal administration this was further proof, too, that politically the miners
needed greater clout and that their interests in Parliament would
be better served by a Labour government.

There was also a recognition among some miners that they
needed their own newspaper to make their voice heard. The lodge
of the Maria pit at Throckley urged the following resolution on
the union:

> Seeing that the press have been fighting against the interests
> of the miners in the late strike, we move that the Council
> Meeting put every effort forward in order to have a paper of
> our own in co-partnership with Durham. (NMA, Minutes,
> 1912, NRO 759/68)

At that same council meeting a resolution was passed advising
branches to co-operate in the setting up of local Labour Party
branches 'so that political propaganda work on Labour Party
lines may be carried out preparatory to contesting Parliamentary

seats'.

The union had, during this period, a distinctively internationalist outlook. They maintained their links, for example, with the
international miners' federation and in 1905 donated £500 to
striking German miners; they supported free trade; they attacked
labour conditions in the mines of South Africa; they criticised
government policies towards China; and they maintained a strong
interest in domestic political issues, particularly in free school
meals and pensions. It is not surprising, therefore, that given
growing competition from Germany in world markets, the acceler-
ation of the naval building programmes and the darkening clouds
of war, the union should speak out against militarism. In 1905
Thomas Burt condemned militarism in one of his monthly circulars,
connecting it explicitly to pauperism:

> Our extravagant national expenditure, much of it on arma-
> ments ... and the heavy taxation falling largely on the prime
> necessaries of life, all must throw increased burdens on the
> poorest of the poor. (November 1905)

And in 1906 the Executive passed a very definite anti-war resolution:

> Recognising that war destroys life and wealth, that it arrests
> the consideration of social politics and the development of
> industry, and that it brings untold misery upon the human
> race, the members of this Association desire to raise their pro-
> test against the expansion of armaments; the encouragement
> of militarism; the loaning of money to belligerents, and the
> tactics of contracting syndicates who for selfish purposes
> promote colonial conquests. (NMA, Executive Minutes, 1906,
> NRO 759/68)

I quote this to stress that through their participation in the
union Northumberland miners were led to an awareness of politics
in such a way that their unionism was not merely politicised but
that it was politicised within an internationalist outlook. The
consequence of this was that in the slide to war thousands of
Northumberland miners could not see Germans, in the abstract,
as the enemy; the enemy was militarism and the economic conditions which supported it. More concretely, however, I mention
the broader political environment because my grandfather always
said that Britain should never go to war with Germany. With
Gordon Brown, the boy he brought up (see Chapter 8), in the
volunteer battalion of the Royal Northumberland Fusiliers and
several of his nephews, although not his own children, of military age, war was something likely to affect him directly.

But this takes the account too far forward. And the form
of exposition itself perhaps communicates too close a connection
between everyday life and the great transformation of politics.
It is important to grasp such connections; but it is equally

true that throughout these years my grandfather experienced his work as a matter of routine. It was simply part of the hum-drum of every day interrupted by the weekends and occasional public holidays. He went to the pit because he had to. His family was growing up and the pit was all he knew. However, the change from the time he was a young man in the pit was, how-ever, great, and the future appeared uncertain.

6 TIME OFF

One way of getting an idea of our fellow-countrymen's miseries is to go and look at their pleasures.

George Eliot, 'Felix Holt'

In contrast to the dark congestion and sour atmosphere of the pit the open countryside suggests freedom and beauty, space in which to breathe and a basic safety in which to relax and to forget. Jack Lawson put it nicely:

It was pay Saturday. I was free to go where I liked and do what I liked all the day. A whole day with the blue sky and fleecy clouds above. Free! Free! If you have risen at four or five on a summer morning and later walked up the village street on your way to the pit, you will understand what that meant. The morning air is exhilarating, the flowers in the little gardens so fragrant, the twittering of the sparrows sweet music – why, the very long commonplace colliery streets seem almost beautiful. And you are going to the pit....
Never does the scent of the flowers possess you, never does the sky seem so beautiful and the birds so much to be envied, as on such a morning. A miner has his compensations. He sees the gloom and he knows grim toil, but he sees the rich, rare morning and he drinks it in. (Lawson, 1949 edn, p. 78)

Jack Lawson, by no means a typical pitman, took his books with him when he went walking – Carlyle, Thomas à Kempis and St Thomas Aquinas. My grandfather never bothered with books but he did seize every chance to walk through the fields and he especially liked walking on clear, frosty nights when he could marvel at the big sky and breathe the crisp clean air. He particularly liked his walks to Heddon; no matter which way he went, along the riverside or through the fields, these were his favourite routes, retracing the steps of his boyhood, perhaps reliving a memory and meeting his old friends.

Walking was for him a special pleasure since, like everybody else in Throckley, he had so little time in which to do it. Until 1938 there were no holidays outside Bank Holidays, Easter and Christmas, and while Saturdays and Sundays were days off from the pit, he still had much to do both at home and on his allotment. Uncle Jim told me that like many of the older pitmen his father did not like holidays; he saw them as days without pay!

Most of my grandfather's waking hours were fully committed between work in the pit and work on the land. In what time was left he relaxed at home; at weekends, depending on how much money he had, he went for a drink. Drinking and gardening brought him into contact with competitive leek growing and flower and vegetable shows. In summer he would, if the chance was there, watch a cricket match, a game he thoroughly enjoyed and one which had been well supported in Heddon under the patronage of Calverly Bewicke, the village squire. It was, indeed, well supported throughout the coalfield before the First World War. The annual visit of the fair, the village picnic, concerts given by the pit brass band or local male voice choirs, perhaps a trip to the coast, were additional opportunities for enjoyment.

Among the things which gave him excitement as well as pleasure were his occasional forays from Lamb's farm with ferrets. He sometimes took Jim ferreting, no doubt recalling some of the illicit nights of his youth poaching on the Bewicke estate at Heddon. But his main preoccupations were not diversions from the daily routines of work; they were part of those routines, and his straining towards self-sufficiency in food, so that his wages were supplemented by his garden, left little time for leisure.

In this chapter I want to show that my grandfather's use of his time off work was constrained by the resources and opportunities of the village and shaped by his own determination to be as independent as possible from the pit; it expressed his basic sense of himself and reflected his priorities, interests and values. His activities away from the pit may have marked him as a miner and member of a particular social class and in so doing indicated something of the 'moral character of a style of life' (see Berger, 1968, p. 28), but they were simultaneously saying much about the man himself.

Tom Burns (1967, p. 742) has argued that 'structures of leisure exist as repositories of meaning, value and reassurance for everyday life'. Seen in this light, Burns argues, leisure cannot be understood simply as a compensation for the deprivations of work. Rather it reflects the way in which people organise their lives, express their autonomy and create meaning and significance for their actions. It is something which must be understood in an historical context too, for the activities which confer meaning and significance in everyday life change through time. My approach to my grandfather's use of his time off work follows that of Burns. In essence this is to treat his actions not just as an aspect of a style of life but as reflecting the structure of the community and society in which he lived. That community was both a setting and a resource.

As a setting it sustained very different interpretations of the meaning and significance of different kinds of leisure activity. Such differences reflected long-standing social divisions in the community. But they also reflected long-standing divisions in society as a whole, for recreational provision in the village and throughout the district reflected, as I shall show, the aims of nineteenth-century social reformers to domesticate the urban working classes. In this way the theme of leisure connects closely with that of class and the changing significance of particular institutions becomes an artefact of changing class relationships. As a resource the community placed real constraints on what men and women could do; low and uncertain incomes from the pit determined that, whatever Throckley people did, it had to be cheap.

How men spent their time was something which had always preoccupied the coal company in Throckley. The school, the chapel, the reading room and the institute, the support they gave to the pit brass band and the prohibition they imposed on keeping dogs were all part of the company's social policy and paternalism. Their opposition to the building of a working men's club and their refusal to allow public houses in the village is particularly noteworthy. Temperance, piety, loyalty and hard work were the values they sought to nurture in their employees. In this respect they were, of course, drawing on powerful images of the responsible employer which were the ideological currency of a whole social class extending far back into nineteenth-century industrialism and social reform.

A good example of this and one highly pertinent to my account of my grandfather is the provision of allotments and gardens in Throckley. The company provided allotments in the belief that gardening both encouraged and sustained 'good workmen'; this was a belief which certainly goes back to the late nineteenth century (see Hammond and Hammond, 1920, pp. 18-19). During that period gardening was encouraged as 'rational recreation' for the poor, having been, during the earlier part of the century, a middle-class hobby emulating the life-style of aristocratic elites (see Constantine, 1979). Such men as Edwin Chadwick, Ebenezer Howard and General Booth of the Salvation Army all held gardening out as a form of recreation to neutralise the lure of the pub. The Throckley coal company did not have to justify their provision of allotments; by the end of the century it was a taken-for-granted mark of a good employer. Having good gardens was something the men themselves also valued and it was, indeed, one of the reasons behind my grandfather's move to Throckley.

But the significance of institutions and resources changes through time and varies among different social groups. The meaning of drinking or gardening to the men of Throckley (or, at least, the majority of them) was not the same as for the coal company. In building their own lives they ascribed to these activities a quite different significance. Their leisure may have been shaped by its context, gradually widening the scope of control they exercised over their own lives. My grandfather's own activities must be seen in this light and the study of those activities becomes, therefore, as I have argued elsewhere, 'the

study of the interpenetration of biography and social structure,
of world building and social control' (Williamson, 1976).

HOME, HEARTH AND PIPE

Apart from going for walks when he could be entirely alone my
grandfather sought his main relaxation in his own home with his
family. Being a family man was a central part of his self-image
and a role he accepted with ease. It was at home that he found
recuperation from work. For him, the house was where he rested.
The family created an enormous amount of work, but the time
he could spare for simply being in the house was time he used
for rest.

He rested by his fireside with his pipe and his paper, and if
the weather was good he sat outside on the bench seat beneath
the window overlooking his garden. He liked being in the house.
The busy-ness of the place, the bairns, the smell of baking
bread, the steady, reassuring tick of the grandfather clock and
the comfort of a well-banked fire pleased him greatly. The whole
scene, including the brass on the fireplace, the leaded range
and the neat mats on the floor symbolised for him what he valued
in his family life - security, success, continuity and cohesion -
for they were the visible product of my grandmother's work and
of his own achievements as a pitman and man.

In that room his worries were few; here he could sit and think
and he always did so with a pipe in his mouth. His pipe was for
him a special pleasure. Apart from a cigar at Christmas or an
occasional cigarette, he smoked nothing but a clay pipe and
then only when he rested. Over the years, quite unconsciously,
his pipe smoking had acquired a ritual character which temporarily insulated him from the clamour of his busy home. Sitting in
his wing-chair facing the fire, he cut his 'baccy' with his knife
and rolled it slowly with a grinding motion in the cupped palms
of his huge hands, all the while gazing thoughtfully into the
fire, totally untouched by the bustle around him. He filled the
pipe carefully and deliberately, building the baccy up in layers,
and lit it with a taper drawn from a box standing in the hearth.
He pressed the curling baccy back into the pipe slowly - all
these actions were slow - and put a small metal lid on top. The
rising smoke was brushed gently aside and, with his knees
crossed and his arms folded, he would sit and stare, his gaze
lost in the blaze of the big open fire, his only movements an
occasional, almost unconscious check on the pipe.

He preferred his clay pipes to be old. He used to leave them
outside in the gutter to age a bit, plugged at one end with a
cleaner and filled at the other with a spoonful of whisky. That,
he used to say, seasoned them and took the dryness off the white
clay. Having watched him smoking many times, I am convinced
that the pipe was of only minor significance; it was the preparations which were important, a kind of ritual as if before prayer,

a way of clearing his mind, slowing it down to the immediate task, contemplation. Years of smoking left that room with a characteristic odour fused into the fibres of curtains, upholstery and floor coverings. It announced powerfully my grandfather's presence in the room.

He read, too. Not novels; he had no interest and even less time. He read his papers, 'Reynolds News', the 'Blaydon Courier' and the 'Evening Chronicle', monthly circulars from the union and anything that was light and took up no real time. Talking was much more enjoyable to him: talking about nothing in particular, reviewing the day's gossip, retailing the news from the pit, listening to the children, thinking through his plans. He helped his children to read, but only if they were stuck; there were no books in the house. His reading was very much a private affair; he didn't read aloud to the rest of the family, he just kept it to himself. In fact, my grandmother used to clear the room for him when he rested, urging the kids out into the street or the wash-house: 'Go on. Get yourselves outside to play while your father has a bit rest,' was what she used to say.

DRINKING AND THE WORKING MEN'S CLUB

At weekends he went for a drink. Drinking was, of course, the chief means of escape from routine for the working classes of Edwardian England (Thompson, 1977), and something my grandfather had done since he was a boy in Heddon. However, drinking for him was not a thoughtless indulgence, expensive or even ruinous to his family. It was not at all like the portraits of drinkers and drinking which filled the pages of the temperance magazines of the time. Quite the opposite: it was a strictly controlled activity, almost solemn, and he took pride in being able to 'take a drink' sensibly, distinguishing himself and his friends from the less respectable boozers in the district. For him, being able to take a drink properly was a small but significant part of his sense of his social status.

He was not a heavy drinker - he could not afford to be - but if the occasion allowed he could drink a lot. Since the coal company would not allow public houses in the village Throckley men before 1908 had to drink in Heddon, Newburn or Walbottle. My grandfather usually drank at Heddon in the Three Tuns, the haunt of his youth. For the sake of his old friends he sometimes went to the White Swan, Bertha's place, just opposite the church, the pub everyone went to when there was a funeral in Heddon, and if he had time he used to call on his way back home at the Frenchman's Arms on the road to Throckley.

In the long afternoons of summer Sundays before the First World War he sometimes took the children and my grandmother to Heddon in his trap, stopping first at Bertha's, then at the Three Tuns and finally at the Frenchman's Arms. He went into

each pub for a drink and bought the horse one, too. He teased
his children with the idea that the horse was so well trained it
knew the time to go on to the next pub for its next drink. My
grandmother never went into the pubs, however; she said she
was too proud for that.

Drinking is a highly ritualised pastime. For my grandfather it
was always something of an occasion to go out for a drink. He
dressed rather stiffly in his best suit, put on his best boots
and always wore his pocket watch and chain and best cloth cap.
On such nights he left the house by the front door and if it
was frosty he would take his silver-topped walking stick with
him. Not for him the silk muffler traditionally worn with a cloth
cap in the pit villages to look smart; he liked to be really smart
and substantially dressed. His clothes announced that out of his
pit gear he was as good as anybody else, and being smart allowed
him, I think, to indulge his own sense of personal dignity without being in any way supercilious. His clothes were a small prop
to his own widely acknowledged respectability. Keeping his
clothes smart, however, as I shall show, was an important element of my grandmother's work.

The Three Tuns was a singing pub which had always been
something of a community centre; everybody joined in and late
on Saturday and Sunday the whole place was filled with lilting,
naughty, sentimental, sometimes crazy, but always eminently
singable songs of the gay nineties drawn straight from the music
halls and made available to millions through cheap sheet music
and pub pianos, or, as in the Three Tuns, a harmonium.

My grandfather always stayed in male company when he was
drinking; he mildly disapproved of women in pubs – an attitude,
I think, which he took from my grandmother and her sense of
what was respectable. It is an attitude which explains his later
preference for drinking in the working men's clubs. After 1908
it was possible for Throckley men to drink in their own social
club. The club was formed in that year and joined the Club and
Institute Union in 1911. The premises for the club were, ironic-
ally, in a street called Stephenson Terrace in a house owned by
George Curwen, the union man and active member of the
Independent Labour Party. He owned the land on which this
property was built and for this reason could circumvent the
social policy of the coal company. My grandfather was one of the
founder members of the club although never a member of its
committee. He was much too busy for that. However, the club
did play an important part in his life and in the village as a
whole, and it illustrates an important feature of social change,
the rise of organised labour extending to ordinary people
greater control of their own lives, through winning control of
institutions designed originally to contain them.

Working men's clubs began in a most inauspicious way in
Victorian England under the guiding hand of a former Unitarian
minister and graduate of London University, Henry Solly. With
the help of Lord Lyttleton and Lord Brougham he managed to

extract moral support and finance from a crop of aristocratic
benefactors to support a movement which was designed to educate
and reform the working classes (see Taylor, 1973, and Jackson,
1968). The Club and Institute Union (CIU) was the child of this
endeavour. It was formed in July 1862. In the manifesto which
Solly drew up to describe the aims of the union it is stated quite
clearly that:

> This Union is formed for the purpose of helping working men
> to establish clubs and institutes where they can meet for
> conversation, business and mental improvement, with the
> means of recreation and refreshment, *free from intoxicating
> drinks*. (Tremlett, 1962, p. 8; original emphasis)

The clubs were seen by Solly and his colleagues as instruments
for education and for the reclamation of drunkards and as pro-
phylactics for the indiscriminate use of beer shops and public
houses.

> The club rooms in every locality will form the strongest
> counteraction to the allurements of the Public House. The
> desire for social enjoyment and the love of excitement are
> the impulses that habitually drive the working classes to
> visit the beer shop. These instincts also form a great tempt-
> ation to reclaimed drunkards. (Tremlett, 1962, p. 10)

It is clear that the idea for such clubs stems, at least in the
case of Solly and Brougham, from a view that earlier attempts
to induce civility and decorum into the working classes had
failed. Brougham had been associated with the mechanics insti-
tutes and felt that their impact had not been sufficient. The
working men's clubs, with their focus on leisure time, might
conceivably be more effective. Without a doubt, therefore, the
CIU was one of many late Victorian devices which attempted to
penetrate the culture of the working classes with a view to
changing it.

The solution to the drink question had been conceived over a
long period of time as the need to find an alternative to the
public house where working men could find sensible and formative recreation. As far back as the 1830s Edwin Chadwick, in
evidence to the Select Committee on Drunkenness, had suggested
that 'public parks and zoos, museums and theatres be provided
to replace the volume of drink being consumed' (quoted by
Tobias, 1967, p. 213). The provision of social clubs for working
men, therefore, was just one part of such long-standing attempts
at social amelioration.

From the early 1870s there was, as John Taylor (1973, p. 17)
put it, a 'revolt against patronage'. Working men's clubs began
increasingly to demand drink, and working men gradually transformed the nature of the clubs themselves although still retain-
ing some of the values of rational recreation and improvement.

For example, B.T. Hall, the national secretary of the CIU,
pointed out in his famous pamphlet, which in the early years of
this century was the bible of the movement, that, in contrast to
the public house where 'the worker must associate with all whom
chance may bring ... the loafer, the blackleg, the soaker and
the rowdy', in the club he would find a selected group of com-
panions and 'congenial company' (Hall, 1908, p. 6). Hall actually
went on to define the central purpose of clubs as 'character
building' and to argue that through 'continuous association, the
constant practice of deference to others, the willing obedience
to self-made rules ... the conscious widening of thought and
habit' clubs directly contribute to the formation and maintenance
of a democratic state. And as if to force the point there were
regular articles throughout the 1920s pointing to the numbers of
clubmen who were elected to public bodies and active in the
wider community. These values were reflected locally in the
leadership of the club and its rules. In Throckley the management committee of the club included many of those well-known
for their work in the union and Labour politics and the co-
operative movement.

The club itself was not in any way political, however. It was
strictly organised around the theme of entertainment. The club
held whist-drives, gardening competitions and weekend concerts.
The main sport was quoits. The club had a quoits square and
held regular competitions and sent its team of players to compete
with other clubs. That streak in the club tradition which in the
1880s had been concerned with politics and education (Taylor,
1973) was exemplified in Throckley by occasional lectures by
the Workers' Educational Association and their annual subscription
to Ruskin College, Oxford. There was a reading room and a
supply of books and one of the committee was librarian. No
record of the holdings of the library exists but I have been told
that Jack London was a popular author.

The club itself was small and decorated like a house. Excessive drinking, bad language, violence and betting were forbidden.
It was a club rule that bookmakers were not allowed on the premises. The club subscribed to the CIU network of convalescent
homes and regularly recommended members to use these facilities.
The club was run by a committee elected to implement the policy
of the annual general meeting. The legitimating rhetoric of this
was a democratic one. An article in the 'CIU Journal' put it
nicely, comparing the club to the state. 'The club', it says, 'may
be taken as the microcosm of the state. It is a perfect model of
self-contained community. In it all men are equal and none hold
position or exercise authority except by the will and pleasure of
his fellows' ('CIU Journal', September 1925, p. 8).

The club, therefore, was free of any damaging association with
drunkenness and the public house; it was a focal point for men
in Throckley with a stable membership of about five hundred.
Being open all day in the period before 1914, it was almost a
community centre, and being a member was a small token of a

man's respectability.

How effective the club was in curtailing the evils of excessive drink is impossible to tell. In the period before the First World War the quarterly reports of the chief constable of Northumberland show clearly that the most common offence the police dealt with was drunkenness. In each of his reports from 1900 to 1916 drunkenness is almost six times more common than the next most serious offences (Chief Constable of Northumberland, Standing Joint Committee Minutes, NRO CC/CM/SJ). This squares with the observations of the German miner who worked in the area in the late 1890s. They make it clear, too, why my grandmother distanced herself from women who took a drink. What Dückershoff said was this: 'Drunkards are as plentiful here, I believe as in Germany. Among women they are more numerous in spite of the many temperance societies.... Tipsy women are as plentiful as tipsy men on Saturday nights' (Dückershoff, 1899). He went on to express the view - accurately, in fact - that drunkenness was the commonest offence the constable had to deal with and that the pawnshop facilitated the habit.

The next most frequent offences were committed against the Education Act, the Gaming Acts and the Highways Acts. We can now only speculate whether these offences were interconnected - whether, for example, the drunkards did not send their children to school, but gambled, and loitered on the highway!

When my grandfather came home from the club he had a cooked supper and, mildly drunk - he was rarely incapably drunk; he was, as an adult, too controlled for that to happen - he indulged his other great pleasure, singing around the piano at home. Left to sing by himself he sang hymns, learned in his childhood and never forgotten, his favourite being Eternal Father, the fisherman's hymn with its special and deeply emotional plea 'for those in peril on the sea'. Aunt Eva says he liked Lead Kindly Light and Abide With Me, too. The weekend sing-song was something of an institution in that house; Jimmy and Ginny, my grandmother's relatives, used to come over from Gateshead. Maggie and Harry were always there, of course, as was Bob, my grandfather's younger brother. When the children were older they stayed up, too.

Throckley, like Heddon, held an annual picnic and flower show. There was the annual visit of the fair or 'hoppings'. At the picnic the pit brass band used to perform and throughout the year there were regular evening performances of male voice choirs of the district, such as Spencer's Gleeming Choir or the Co-operative Choral Society. During the summer there were frequent parades in support of various charities such as homes for aged miners and the Throckley Colliery Sick and Accident Fund. Throckley had its football team (Throckley Villa) and cricket team. Events such as these offered some chance of entertainment and diversion. My grandfather always used to go to the picnics and as often as he could to performances of the band. He helped Tommy Lamb, a local farmer, to exhibit his cows at

the picnic, and since, on the day, the cows still needed milking
my grandfather used to milk them, keeping the milk for his own
use.

On the periphery of his interests, although, of course, central
to those of other people, there was a good deal of pigeon fancy-
ing and, for the younger men, a cycling club organised by Billy
King, activist in the co-operative store. My grandfather took a
mild interest in pigeons since his brother Bob, just up the
street, was a keen 'fancier'. Like many pitmen he kept an eye
on the skies; flights of pigeons indicated who were in the gardens
and what they were doing and pigeon talk with its uniquely
arcane jargon was a stable feature of conversation among the
men.

Then there were the chapels, the church and much more
cerebral activities like the Throckley Home Reading Union Circle
organised, before 1914, by Mr Heslop.

There were illicit gambling schools up the back lonnen. My
grandfather avoided anything to do with gambling, but he did
encourage his children to involve themselves in the church and
the chapels. Indeed the only moral tale my grandfather told his
children concerned gambling. He told them that he had once
been persuaded while working at Heddon pit to put his week's
wages on a horse. The horse, of course, lost, and he used this
cautionary tale to warn his sons off betting.

GARDENING

Beyond seasonal diversions and the purely personal indulgence
of a weekend drink my grandfather's time was committed to his
allotments and the mild rivalries of showing vegetables which
gardening spawned. And this commitment to gardening, arising
from wholly utilitarian motives, indicates a great deal, not only
about my grandfather's own personality, particularly his deep-
rooted determination to be independent of the pit, but also of
the social organisation of the community as a whole and the
values embedded in it.

My grandfather's work in his gardens and the high priority
which he gave to it displays all the features of what Becker has
called 'commitment' (Becker, 1970). Firstly, he chose to spend
a great deal of time gardening. Secondly, his gardening had
important consequences for many of his other interests. It directly affected the standard of living of his family but it was also
part of his own self-respect as a good workman and a man and
of his relationships with his friends in the village. These features
are what Becker would call 'side bets', i.e. interests or actions
which have become part of the original commitment. To give up
gardening would have been expensive both in real economic
terms and in terms of the quality and kind of social recognition
he was accorded in the village. His gardening, paradoxically,
was part of those complex forces which kept him a pitman. It

was one factor which made him unwilling to countenance moving
out of the pits or even to seek a council house. These 'side bets'
(i.e. gardening), therefore, helped to shape his priorities and
his self-respect. It would trivialise the significance of gardening for him if it were to be thought of simply as his relaxation
or a contrast to his work.

Until the early 1930s my grandfather rented three allotments
which he kept cultivated in addition to the long front garden
of the house. Until 1914 he kept a horse, a trap and a flat cart,
and in the gardens he kept hens, geese, sometimes ducks,
rabbits and pigs. He did not keep any cattle because it was not
necessary. He worked as often as he could on Lamb's farm and
was paid for that work largely in kind. He grew a full range of
vegetables, potatoes in particular, storing them over the winter
in his garden shed. Because of his heavy commitments in the
garden and with the animals he tried to arrange it that he always
worked the foreshift from 3.30 a.m. to 11.30 a.m. This gave him
most of the afternoon and the early part of the evening to get
through his chores.

Bill says of his father that 'he was almost a bloody farmer',
and my grandmother often, knowing full well the impossibility of
it, used to say to him, 'Jim, you should have been a farmer.'
The benefits for the family of his gardening were obvious. Jim
told me:

> I've known me father go to the garden and fetch the whole
> dinner in. Howk the tetties, cabbage, and kill a hen. That
> was our dinner. It didn't cost a ha'penny, just out of the
> garden. We had about two or three chains.

But he did enjoy this work. He enjoyed working with and caring
for animals. Not that he was sentimental about it. He could pull
a chicken or break a rabbit's neck with clinical precision and
without regret. Cruelty was another matter; he would not
tolerate that. He particularly enjoyed work with horses. This
was, of course, something he had done since boyhood and during
his days as a driver in Heddon pit. Feeding, grooming, cleaning
and exercising his horse took a great deal of his time. Talk of
horses prompted uncle Bill to recall a very early memory of his
childhood. Bill remembers playing in deep snow at the bottom of
the street in the winter before the First World War, and seeing
his father leading his horse to the 'three cornered field' opposite
Mount Pleasant. When Bill asked him why he had the horse trot-
ting hard in the snow he got the crisp answer, with no further
explanation: 'It's good for its fetlocks.' 'That', said Bill, 'was
me father all over. Always bloody busy.'

Garden work forced him into regular routines and brought with
it various sorts of obligations and commitments to other people
which he repaid in kind. It also involved his family, especially
the boys, and brought him into direct contact with the co-operative store since he used to sell his pigs to the butcher's depart-

ment there. Garden work must therefore be understood in its
own context just like work in the pit; it had its own constraints
and opportunities and it, too, over the years, helped to shape
my grandfather's whole outlook.

KEEPING PIGS

His pig keeping illustrates these points very well. Shortly after
moving to Throckley he started to keep pigs. The equipment was
simple. His wash-house shed housed the boiler to heat the pig
swill and was at the same time a place where the slaughtered
carcass could be jointed and temporarily stored. The pig cree
(pen) in the garden was a simple cement structure which he
built himself. His experience both at home as a boy when his
father kept pigs, and on Law's farm at Heddon, served him well.
He knew about pigs and was confident about keeping them.
   Feeding pigs is a problem; my grandfather solved it by regularly collecting waste food - peelings, cabbage leaves and the
like - from neighbours and acquaintances. When his sons grew
strong enough collecting the 'peelins' was one of their key tasks.
Bill explained:

> We used to come home from school, this is true, and get the
> barrow and go away to Newburn and Walbottle to seek peel-
> ings.
>   We had a barrow the size of a settee with great big iron
> mangle wheels on it. One or two kept pigs but not like us. Me
> father was nearly a farmer. Pigs and hens. The pan was
> never off the fire in our house, a great big pan. As soon as
> me mother took the kettle off the fire me father used to put
> the big pan on, boiling for the pigs. Me mother used to go
> mad. I've seen us kids pinch the tetties he'd boiled, too.

Jim explained that he kept four or five pigs at a time. He used
to kill a couple for bacon, one about Christmas, the other in
March, and fatten up two or three others to sell to the co-
operative butcher. The pigs brought him in a bit of cash, but
better than that they gave the family an almost continuous
supply of ham; the salted hams used to hang on meat-hooks in
the cool of the stair bottom.
   Pigs are extremely productive. They give a high-grade nitro-
genous fertiliser. At pig-killing time black pudding (made from
pigs' blood), sausage and broth were in plentiful supply, so
much so that a lot of it could be given away to those neighbours
who had put up with the inconvenience of storing their waste
food. Pig killing was a great social occasion in Mount Pleasant.
Mrs Allen used to come in and help make the sausage and prepare
'pig's cheek' from the pig's head, and potted meat. The children
looked forward to it because they could have the pig's bladder
which, dried and inflated, became a very useful toy, a cross

between a balloon and a football.

For my grandmother, pig killing brought conflicting emotions. Involved occasionally with the feeding of the pigs, she could not stop herself thinking of them as pets. So that the children should not hear the frenzied squealing of the pig as it was struck by the heavy mell of the store butcher, she made certain that they were always well out of the way, usually at school, and she herself had as little to do with the actual killing as possible.

Unlike the work my grandfather did with his vegetables, which was seasonal, pigs had to be fed all the year round, and in winter he used to carry his pails of swill down Mount Pleasant and along to his garden with his children carrying hurricane lamps to lead the way. In fact, my mother told me, he often had to work in the garden by the light of his lamp, a chore, she emphasised, which presented him with few problems since he was very used to work in the dark! But the payoff was worth the effort: a secure supply of food and a bit of cash, too - a small contribution to the independence he strived for. It was an independence, however, which incurred obligations and which presupposed a network of friends who helped him to feed the animals. The successful keeping of pigs measures my grandfather's success in the community.

The problems of the pigs and his garden framed his sense of what was urgent and important and what his priorities were, and they certainly left him little time for other kinds of commitment. The time scales which preoccupied him, apart from those of the pit, were the relatively short ones of pig fattening, planting and harvesting and these translated themselves into daily commitments and the need for predictable routines.

LEEK GROWING

Gardening brought him into a mild rivalry with other gardeners through the leek and vegetable shows which took place every year in almost every village throughout the coalfield. My grandfather's leek growing was the only non-utilitarian aspect of his gardening and he spent a lot of time on it, as did many other pitmen. Like keeping pigs, leek growing is something of an institution which can only be grasped in its social context.

The basis of competitive leek growing is simple; gardeners competed with one another to grow the longest and fattest leeks - in the period before the First World War they were mainly interested in the length - and the winners received a simple prize. Growing huge vegetables for show was a popular pastime throughout the district, and in 1902 Dr Messer, the medical officer of health, was moved to comment on the fact. Rising to a fine irony and writing explicitly about the lack in the district of public provision of play space for children he noted:

Herbert Spencer says that had Gulliver narrated to the
Lilliputians that the men vied with each other in learning
how best to rear the offspring of other creatures and were
careless of how best to rear their own offspring, he would
have paralleled any of the absurdities he relates to them.
(MOH, Newburn UDC, Annual Reports)

For the growers, however, there was no contradiction in what
they did; leek growing was a mild form of rivalry which thrived
on and presupposed friendship among men.

Lying behind this rivalry was, and still is, a rather complex
social network which regulated the distribution of good 'breeds'
of leek and a body of knowledge of how to grow big leeks which
is so exclusive and esoteric that the uninitiated could never
hope to succeed in winning shows (Williamson, 1973). Leek growing was organised from the public houses and, after the First
World War, from working men's clubs. It was an exclusively
male activity financed by weekly lotteries and it was deadly
serious. Before 1914 the prizes were simple. In 1905, for example,
the first prize at the Frenchman's Arms, a show my grandfather
attended, was a blanket. The prizes were, and still are, invariably something for the home; this was a sop to the wives which
justified the amount of time the husband spent in the garden
tending his leeks. Each grower had his own views on what would
make the leeks grow big; in this respect it was a secretive
pastime. However, families did have an opportunity to become
part of it. It was traditional for there to be a leek supper after
the show for which broth would be made and distributed freely
to everyone, and the women and children could come into the
pubs and clubs to see the display of leeks. For the young boys
this was often an occasion for an illicit sip of their fathers' beer
and for the wives a chance to pour mild but jovial scorn on their
husbands' horticultural skills.

My grandfather always grew leeks in his front garden; they
were safer there, and the only potent for growing big leeks that
he swore by was the scrapings of a baby's nappy. Like all leek
growers, he enjoyed the elliptical discussions about how to grow
leeks which always failed to extract from 'the big growers' just
what their secrets really were. In this sense, what was important
about growing leeks was not the leeks themselves - indeed, they
were often so force-fed as to be inedible - but the social contacts
of the sport which gave the drudge of gardening a mild competitive edge of fun.

My grandfather was an active competitor but not a successful
one. Old Tom Stobbart says he was not a patch on himself at
growing leeks. It is hardly surprising, for successful leek growing takes an enormous amount of time, the one thing my grandfather did not have. For in addition to his garden and his job
he had his family and his repairs to see to. He looked after his
horse; he built his own hen crees and sheds; he cobbled the
family shoes in the wash-house; he cut a great deal of his own

grass for hay, and even he needed rest and sleep!

Where he was really successful, I think, was in so arranging his time and his activities that his hobbies were absorbed in his work and there was no inconsistency between what he did and his family life. When his children talk about their relationship with him they always do so in terms of such things as helping in the garden, riding in the trap, caring for the animals, sorting potatoes in the shed and being dispatched outside when he needed rest. He absorbed his kids into his hobbies effortlessly and in so doing taught them how to work.

In 'Coal is Our Life' the leisure of miners is described as 'vigorous' and 'predominantly frivolous', a word which the authors use 'in the sense of "giving no thought for the morrow"' (Dennis et al., 1956, p. 130). This is a claim made often enough about working-class leisure as a whole (see Meacham, 1977). Dennis and his colleagues attribute the frivolity of it to the insecurity of pit work which encourages miners to 'live only for today'.

Such an interpretation makes little sense of how my grandfather spent his time. Their general theme that the solidarity of the men underground is reinforced by their social relationships on the surface, fits the case well enough. My grandfather was totally absorbed in a male world where helping, sharing, working hard, being strong and upright were all heavily reinforced social values. And the busy-ness of it, I am sure, kept him so preoccupied with the flow of everyday details that he rarely stood back to take a critical look at his job and his conditions. As with his work, there was little in his leisure to produce a 'reflective awareness' which would have distanced him from the taken-for-grantedness of his daily life to enable him to see its limitations. He was far too preoccupied with what Richard Hoggart (1957, p. 87) has called 'the concrete' and 'the personal' to be reflective. By not contradicting the dominant assumptions about work or the standards of respectability which prevailed in the village, his leisure actually reinforced his work status as a pitman (cf. Dennis et al., 1956). But he was no mere victim of social control. He also found in his leisure, as I have shown, autonomy, self-respect, some independence and greater security, qualities of his life and social values his work alone could never have confirmed.

So far as standing back from his position is concerned, he was very different from my grandmother. Living in the shadows of his life, she did the worrying and bore the psychological risks of his working underground, which underlines heavily, as I shall show in the next chapter, that the costs of exploitation are not always borne where they are incurred. The taken-for-grantedness of his daily life was almost total and he was extremely content.

But it could not be said that his leisure was frivolous, and he was by no means unique. His 'leisure' was purposive and useful and through it he realised a basic self-respect and gave his family greater security.

Dennis and his colleagues were not wrong in their judgment, however; they were merely describing a mining community fifty years after the time I have been concerned with, in a society-wide framework of social security which, while hardly adequate, was none the less politically light-years from the one my grandfather had to cope with.

7 DOMESTIC WORK

> There was a big discrepancy, when I was a boy,
> between the collier who saw, at the best, only a few
> hours of daylight - often no daylight at all during the
> winter weeks - and the collier's wife, who had all day
> to herself when the man was down the pit. The great
> fallacy is, to pity the man.
> D.H. Lawrence, 'Nottingham and the Mining Country'

Without women mining communities would not exist; they would
be labour camps. Yet the history of mining communities has been
distorted by an almost exclusive emphasis on pit work, although
pit work in the form we know it in the nineteenth and the early
twentieth centuries was only possible because of the way in
which women worked in the home. Housework was as central to
the winning of coal as the graft of the miner underground. It
was through housework that the miner could be prepared day in
and day out to return to his work. It was the home which supplied the pitman with his main rationale for working. And it was
the home, in a cruelly ironic way, which tightened those bonds
between families and pits which tied the miner's children to their
father's way of life, seeking in their own right work underground.

Just as the social structure of the industry defines the class
position of the miner the organisation of pit production defined
the family position of the miner's wife. What the miner's wife
was able to achieve in her own life, what satisfactions she
found and what hopes she could nourish, were all limited by her
own expectations born of her education, the predictable routines
of the pit, the resources of the village and the attitudes and
expectations of the men. Lacking the kind of upbringing which
in the men developed a devil-may-care attitude to the pit or a
fatalistic tolerance of its risks, and insulated from that supportive framework of workmates and pit talk, unable to find
employment in their own right, the women were left to bear the
psychological risks of the pit and the precariousness of their
own and their children's security. The only way to avoid the
anxiety their condition produced was to be immersed in routines,
not to think too far ahead and to value immediate blessings (see
Davidoff, 1974). And with small incomes - an almost total dependence, in fact, on a fortnightly wage - the immediate blessings

were not material ones but concerned the quality of personal
relationships. And here, I think, lies the key to the almost total
preoccupation of miners' wives with the unfolding patterns of
family life, with their children, grandchildren, and their husbands and their far-flung kin. And here, too, is the source of
the nostalgic mother–centred family image of the past which
anyone studying such communities encounters so frequently. In
this chapter I shall examine only the work of the house; relations
in the family are discussed in the following chapter.

HOUSEWORK AND ROUTINE

Viewed as work, housework for my grandmother had a number of
characteristics (see Oakley, 1977). It was physically hard,
relentless, monotonous and largely unacknowledged as requiring
any special aptitude. In addition it was considered, both in the
family and by those around the village, to be properly woman's
work, a fact she herself also accepted (see Benney, 1946, and
Dennis et al., 1956). It gave her little freedom of choice about
how she could use her time and it was extremely repetitive. It
was carried out in a very confined space and brought her into
contact with very few people. Her responsibility for planning
the family budget was not made easy by the unpredictable size
of the fortnightly pay, so there was always an underlying
uncertainty to her work which created worries for which there
was no easy solution. Nor was there ever any real prospect that
the work might change in significant ways. My grandmother was
basically a very contented woman, but in the circumstances no
other attitude would have been psychologically tolerable. She was
in any case well prepared for it. Her education had led her to
expect nothing more and her own upbringing in a mining area of
Durham had taught her what she needed to know.
    She acknowledged her 'blessings' readily and found great
comfort in small things, a quiet hour with her magazines, tea
with a neighbour, finishing a piece of clothing and so on. She
rarely complained; she simply, as she used to say, 'got on with
it', rendering the acceptance of her lot a kind of value in itself.
In any case, because of the acknowledged fact that men worked
hard in the pit, there were few ways in which women could
legitimately claim that their work was especially onerous (Dennis
et al., 1956). Her real task, as she saw it, was to create comfort-
able conditions for her husband, not because he in some patri-
archal way demanded this but because she thought that this was
where her duty lay. This attitude of acceptance and forbearance
does not signify in her an uncritical view of her social position;
she was, in fact, acutely aware of the limitations of her life in
Throckley. Rather it was a way of making her life bearable, of
adjusting her expectations to resources and obligations so that
her satisfactions, such as they were, could actually be realised.
The qualities of patience, forbearance and contentment which

she exemplified are not, therefore, simply aspects of her personality; they have to be seen as personal qualities shaped considerably by the work she was compelled to do. Her work, in
short, must be seen in the appropriate historical context.

She was not a passive victim of circumstance. She brought to
her work her own unique style of commitment and found in it
considerable satisfaction, and the key to understanding that lies
in the way, through the strict control of her work, she created
spaces in which she could be free of it and through that freedom
find personal autonomy (see Davidoff, 1974). Here, then, is the
paradox of routine; through ordered routines she found her
freedoms.

The setting for her work was the house itself. Inside they had
one main living-room with a bed alcove and a black-leaded fire
range and oven. Upstairs there were two bedrooms. The kitchen
was at the back of the house, its only real facility a sink with
a cold-water tap. To the left of the living-room window, beside
the settee, stood the grandfather clock aloof from the room.
Against the alcove wall stood a piano. Opposite the window was
a sideboard. Right in the middle of the room, caught in the light
of the hanging gas mantle, was the table, covered with an oilcloth and surrounded by chairs. The fourth wall was dominated
by the big black range and brass fender. The floors were
covered in oilcloth and 'clippy mats'. The whole room was hopelessly congested. When the fire was 'bleezing' it was suffocat-
ingly hot and even in summer the fire was kept on for cooking.

The work done in the house followed directly the routine of
the pit and its different shifts. My grandmother brought some
order into her tasks by evolving certain routines and a weekly
division of labour. These routines were vital to her; through
them she achieved control of her work and some freedom. Precisely what these routines were has been fully described to me
in writing by my mother and I quote this document at length.
Firstly, the weekly routines:

> 'Monday' was always called 'Cobbler's Monday.' Why? I don't
> know. Mother made our breakfasts - no such things as cereals,
> or fruit juices - the main course was porridge or bread boiley
> (that is bread cut into cubes, swollen with boiling water, then
> strained off and boiling milk poured over and sweetened with
> sugar, lovely it was) then bacon sandwich or a boiled egg and
> toast, toasted in front of a big fire, no fancy toasters, then
> off to school.

My mother is here recalling the period of the First World War and
the early 1920s.

> Mother got cracking then. First she brushed my dad's Sunday
> clothes and the family's, folded them and put them away for
> the next weekend. The dinner on a Monday was always called
> 'cold warmed up', everything left off Sunday dinner was fried

up and served on Monday. Tea was cakes and scones that were cooked on Sunday, supper sometimes a kipper or a few chips or an egg or anything left over from the weekend.

The bread bin was never empty, mother baked all her own bread, tea cakes (stotty cakes). They were lovely when they were first made. No such thing as calories, no one minded being fat. My mother used to say, 'The thicker the meat the stronger the man', and 'Pack the food into the bairns when they are small and they grow up to work for all the luxuries later in life.' Mother did not do much housework on a Monday. I think that is why she said 'Cobbler's Monday'. The house was nice and clean after the weekend clean up and not much cooking. She kept Monday to catch up with the mending of garments, darning socks, making new clothes out of old ones. She made our clothes. She knitted all the socks and stockings. She made my dad's shirts, cut up the old ones and made small ones for my brothers. Dad repaired all our shoes.

Monday, then, left my grandmother with a little more time than usual, there being little cooking to do. It also gave her a chance to make clothes, a task, while born of necessity, which she none the less enjoyed. Tuesday was washing day:

Mother always made a big pan of broth. That was prepared on Monday. All the produce was grown in our own garden, except the barley. So broth on a wash day was handy. She used to say, 'It is just ready for the floaters to go in.' (Floaters was dumplings.) After the broth and floaters we got meat sliced, served with potatoes or sandwiched in stotty cake, and tea. After a meal like that we were all satisfied. So tea was not important, bread, home-made jam, or cheese was all we needed. Supper again, a patch up, but always tasty and good. My mother always said, 'Never go to bed hungry.' We always got something.

It is noteworthy that in her written account of washing day my mother mentioned nothing of washing itself or the disruption to the house which it brought (a disruption celebrated in the famous Tyneside song, Weshin' Day, with its opening lines, 'Of all the plagues a poor man meets A-lang life's weary way, Thor's nyen amang them aall that beats a rainy weshin' day').

Washing was done in a poss tub (or wash tub; the action of using the washing stick - the poss - was known as 'possing'). 'We used to double poss. One on either side; one went up the other went down.' My mother then explained that the poss tub itself was acquired from a local fish and chip shop. The shop used to buy its frying fat in big barrels which, washed out and scrubbed clean, were ideal. But they were not just for possing. My mother says that she had many a bath in the poss tub. My grandfather used to fill it up with hot water, stand it in the kitchen, draw the curtains and send the girls in to get their

bath. They came out and dried themselves in front of the fire in
the main room. 'I used to love it; nice and warm in front of a
blazing fire.'

Washing day involved a great deal of inconvenience. The water
had to be carried from the one cold tap in the kitchen and heated
in the outhouse boiler and on the kitchen fire. The fires had to
be kept going and the dirty water poured down the gutter in
the back lane. On dark, damp days the room was steamy and
congested, and if the clothes could not be dried on the garden
line they had to hang in the shed and in the living-room, detract-
ing from the basic comfort of the room.

'Wednesday', my mother says, 'was another busy day.'

Piles of ironing. Clothes were a big heavy task. Mother always
got the coal oven going, it helped to air the clothes after they
got ironed and hung over the big brass line. So the oven had
to be put in use. Usually that day, she would make a great
big taty-pot (potatoes, onion, veg, a few roasted dumplings)
that was working while she was ironing. The coal fire had to
be kept bright for the iron flat irons. After the hot pot came
out of the oven a dish full of herrings was popped in. Mother
always said, 'Never eat herrings until they flow through May
waters, never eat rabbits until there is an R in the month,
same with pork.' So the diet was always watched.

Still dealing with Wednesday and ironing, my mother's account
then moves on to an entirely new theme, although in her reconstruction of the events the issues are clearly connected. And
what it illustrates, I think, is this: it is quite impossible to clas-
sify the separate components of housework as independent tasks.
Housework has its many layers and different tasks are done
simultaneously. The new theme in my mother's account was mat-
making and knitting.

Mother always had a mat on the frames, so if she had a spare
few minutes at all, down to the mat she went. She always had
socks on the needles. She kept her knitting behind the piano
lid, if she sat one minute out came the knitting.

I shall have more to say about mat making later since this particular activity, quite apart from its intrinsic importance to the
household, was a focal point for neighbourly contact.

'Thursday was a full baking day, bread, tea cakes, a stone of
flour at a time, tarts, meat tarts, pies.' The cooked food was
kept in the cold pantry and, although great care was needed in
the summer to see that food was all right, little went off. There
were plenty of eggs from the chickens. There was always bacon
from the pigs my grandfather kept. The basic materials for baking were in good supply. The work, however, was hard and
skilful because any cooking involved carrying coal, and the oven
temperature was difficult to regulate.

Friday, like the others, was 'a very busy day', but it was one
they looked forward to in that it marked the end of the week and
Friday evening was an important social occasion for the whole
family:

> In fact, every day was hard as our house was an old-fashioned
> colliery house, a big open fire, coal oven, water heated at the
> side of the fire place (which held about four pails of hot water),
> but Friday was a hard day. The fireplace was all cleaned out,
> reset, bars and stove black-leaded – that was put on and
> brushed off to make it shine. The fire irons and fender were
> all polished with Brasso, the big brass line polished, the
> hearth whitened with a brush – it was sparkling when finished.
> The plush mantel mirror around the top, that was the pride of
> the house. Then the floor washed, no carpets just oilcloth on
> the floor; then down went the lovely mats, the furniture
> polished with a wash leather, washed in vinegar and water,
> clean covers put on. By a Friday night our house was spark-
> ling. For supper, out came the pies that had been made on
> Thursday, pease-pudding, home-made black pudding if a pig
> had been killed – Mam made it herself – or else we kids called
> at the killing shop at the Co-op and got a can of blood, she
> made the black pudding from that.

Friday, then, was a day of preparation. After the retirement of
my grandfather it was also the main social night of the week,
but during the time he worked it was a mere herald to Saturday,
pay day and the last day of the working week.

> Saturday, well, the house was nice, so it was a day Mam
> could relax. Her only luxury was her weekly Welcome Book
> (2d) and her knitting. She liked us bringing our friends.
> Saturday the lads went out. Dad at the pub. By then I had
> learned to play the piano, so I brought my friends home, boys
> and girls, we had a nice sing-song, supper and a chat. By
> then my Dad came in from the club. He loved to see us happy
> and singing.
>   I always remember my Mam saying on a Saturday night, 'I
> will sleep tonight ... because none of them are down the pit.'
> Those words played on my young brain. I used to say, 'Mam,
> I will never marry a pitman if it causes so much worry.'

My mother then turns to another theme which, while not directly
connected with domestic work, none the less bears directly on
the psychological costs of being a pitman's wife:

> My mother watched the clock and kept going to the door to
> watch for the lads and my Dad coming down the street and
> she always said, 'Thank God, here they come,' and out came
> the dinner. That stuck in my mind as a child. I was scared
> for my mother.

A further aspect of this underlines just how far my grandmother's life was lived in the shadow of the pit. Like other wives
in Throckley she was extremely attentive to outside noises. The
bustle of the street, the crunch of hobnailed boots, children
playing, the calls of the hawkers, the intermittent clanking of
tubs from the dilly line, the throb of the pumping engines of the
pit all implied normality. But an unexpected blow from the pit
buzzer could penetrate the normal noise immediately, spreading
alarm and anxiety through every house. Saturday and Sunday
were the only days when she was free of that subliminal threat;
only then could she feel she was safe.

Saturday evening was clearly a high spot; apart from my
grandmother and my mother everybody went out. After a busy
week, my mother used to like to stay in.

Saturday, the night I loved, sitting in with my mother, our
two selves. The lads went out, my Dad was at the club, and
above all, our Eva was out with her kids – she always visited
her mother-in-law on Saturdays. Mother always sat and knit,
she loved to listen to me playing the piano.

Sunday finished the weekend; it was another day of cooking and
cleaning although only for the women:

We all helped in the house, I should say, the girls. One made
the beds, one washed dishes, one cleaned up the house. Mam
cooked the dinner, made cakes, tarts, scones, so again we
had a full table. At night, again our friends came home after
church, again the piano played. As always Dad came home
from the pub and at once he sang; he loved the piano and he
loved his family.

Sunday morning was a regular time for a very serious ritual.
This was when the grandfather clock was wound up and set, a
task for my grandfather alone. No one else dared touch the
clock. He used to give the mechanism a brush with a cockerel
feather and carefully wind up the clock. His private name for it
– and this was something of a family joke – was Hannah. The
Sunday ritual with Hannah often prompted my grandmother to
exclaim, 'You think more of that clock than you do of me.' And
she told him more than once that she 'would have his coffin made
from it'. The clock was his prize possession and I can remember
vividly as a small child feeling awestruck by this ominous old
thing, hardly daring to go near it, an attitude passed on to me
by my mother. It was exactly the same for her and all the other
children in Mount Pleasant.

The house was always untidy again by the end of Sunday night.
The important things for the next day were, however, always
ready. Pit boots had been dubbined, pit clothes were ready and
dry, and there was plenty of food for the baits.

My mother's account of housework covers a wide range of

different tasks so some small elaboration is necessary. 'Cleaning up the house' involved dusting, polishing, washing floors, scrubbing, and everything was done with simple equipment. My mother explains:

> No hoovers, just brooms, brushes, buckets of water and scrubbing brushes, wash leathers, elbow grease. Outside steps were washed every week and sandy stone rubbed on them. We got sandy stone off the ragman for old rags. Mother was choosy with that. She liked soft stone. It had to stand in water first, then be rubbed on the steps. Rain did not wash it off but rain washed chalk off so sandy stone was the stuff. In later years white paint did the job but everyone down Mount Pleasant liked their steps bright with sandy stone.
>
> Also the windows was the pride and joy, nice lace curtains. Mother was lucky to have a niece in the second-hand trade so aunt Maggie and mother got the best curtains and our windows were the envy of lots of neighbours.
>
> My mother used to say, 'People knew the inside was clean when the outside was good.'

A vital part of the outside were the middens and the gutter. My grandmother, like her neighbours, made sure the main open drainage gutter was kept clean and well weeded and that the midden and toilet were kept smart. Both were in a real sense public; the family got to them by crossing the back lane. At night they had to use a candle or my grandfather's pit lamp. And as if to make a virtue out of necessity my grandfather always argued that inside lavatories were not healthy. Since these were areas where the children played - of which more later - they had to be kept clean.

It seems hard to avoid the view that my grandmother's life was dominated by a kind of 'compulsive domesticity' (see Rosser and Harris, 1965). She did little else but work; housework was her life. But through housework she found a basis for her own self-respect and the recognition of others. And this really does need emphasis. Many have noted the uniformities of mining communities - men had much the same wages; they lived in similar houses; they knew each other's affairs intimately (see, for example, Dennis et al., 1956; Benney, 1946). Any outward sign of a claim to be different, particularly a claim to be in any way superior, was looked down upon and would invite gossip and adverse comment. Phrases like 'she's got a nose above her mouth' and 'she's getting above herself' or derogatory words like 'posh', 'la-di-da' and the like could be savagely applied, bringing in their tow the social isolation which would make life impossible in a community which depended so much on mutual co-operation. Any claim to a special self-respect had to be based, therefore, on an obvious claim to possess in some acknowledgeable way those qualities which all women in Throckley possessed, but to do so in such a way that no insult to others or feelings of superiority over others were implied.

My grandmother found that self-respect in her diligent housework and the visible signs of that - the bright windows, the
sanded steps, the line of white washing, the well turned out
kids - each a simple yet powerful symbol of a personal dignity
which much in her environment threatened to destroy, and each
amply compensating in its symbolic force for the deficiencies of
its material worth.

Within the weekly routines that have been described there
were the daily tasks - making up the feather beds, tidying away
clothes and, as if to seal the connection between her and the
pit, making ready for the men to return home from work. Water
had to be heated ready for their bath and every day my grandmother had to make sure there was some fruit loaf in the house,
because when my grandfather came from the pit he liked a slice
of fruit loaf - eating that before anything else, even before
taking his bath. Food eaten in the pit is said - and was said -
to have a distinctive taste. Indeed, my mother occasionally
arranged for my grandfather to take bread and sugar sandwiches
down the pit with him only to bring them up again for the children to eat. The pit was thought in some way to add a special
flavour to the bread. But it was perhaps to get the taste of the
pit out of his mouth that my grandfather had the fruit loaf - in
curious contrast to his womenfolk who, perhaps unconsciously,
were seeking some closer contact with the underground world
from which in every other respect they were totally excluded.

The time before his return was an anxious one. But routine
took away the sting. His bath was always ready, and after his
clothes had been dadded on the outside wall and put in the oven
to dry he took his bath in front of the fire and was followed, as
they grew older, by his sons, all using the same water. And as
if to make a virtue of yet another necessity my grandmother used
a phrase which my mother used to use when my brother and I
complained of the need to use the same water, 'dirty water washes
cleanest' - the perfect rationalisation!

Despite the hectic housework, the pit pressed in on the home.
Damp boots cluttered the small passage from the back door and,
as my mother once told me, almost despairingly, 'Our house was
never tidy, you know, always pit clothes around the place.'

Spring was the time for a huge clean out and for redecoration.
Wallpapering and painting were done by my grandmother although,
as the girls got older, they helped too. And spring cleaning was
the occasion to lay new mats. The neighbours helped with the
mats. A casual visit for a cup of tea would often be the occasion
for friends to sit down and work at the mat for a few minutes,
and kids, of course, could always fill up a little time cutting
clippings.

The resources of the village - its shops, its water supply,
its transport and its services - served to fashion domestic work
in particular ways. Most of my grandmother's shopping, for
instance, was done in the co-operative store, there being few

alternatives. But shopping or, rather, purchasing goods was
not confined to the Throckley shops themselves. There was a
considerable traffic of hawkers and cart traders up and down
the pit rows from whom people could buy a wide range of goods.
These added to the limited variety of the local shops. And they
added a good deal of colour to the drab streets. Shopping,
however, had to be carefully planned and budgeted for and the
main responsibility for this was my grandmother's. Here again
the women bore the burden of the uncertainty of the fortnightly
wage and the credit they needed to manage. My mother explains:
'All the people in our village had to purchase clothes, footwear,
bedding, on weekly payments. The Co-op was the main stores,
but sometimes a bit pricey.' But she goes on to show how street
traders offered some alternatives:

> I remember a traveller coming to our house, a Mr Soloman.
> My Dad liked good suits. He was tall and in no way could he
> get a suit ready made, so a good tailor had to do the job.
> When Soloman left, another man called Mr James took over.
> After he left Walter Proudlock started a workroom in a hut
> behind Curwen's garage. He canvassed and got orders. By
> then my brothers wanted good tailored suits, so we were
> considered good customers, a regular big payment each week.
> He was the tailor for years after. I also remember a chap
> called Mr Green. He travelled the street every Friday. He had
> a huge cart driven by a big draught horse. He always had
> two big hurricane lamps hanging each side of his cart. Also
> he rang a big heavy bell and he always shouted, 'Lamp oil.'
> He sold gallons of paraffin but also hardware, tins, pots and
> pans; in fact, everybody said that Green sold everything
> from a pin to an elephant.
> Jack Miller, he lived beside Throckley school, he used to
> go around in a little tub cart, pulled by a horse. He sold
> farm butter and fresh eggs. His face was red and weather
> beaten but his butter and eggs were good. Mother liked his
> butter; we had our own eggs.
> Another man called Mr Bone used to come around with a
> basket selling bread, tea cakes, muffins and crumpets. He
> had a shop where the chemist is at Throckley. His daughter
> served in the shop. They were two crabby old people. He
> never sold much in Mount Pleasant, as everyone made their
> own bread.

My grandmother had to deal with these people; to know the times
of their rounds and to be ready, if necessary, to drop what she
was doing to go out, usually among others, to buy things from
the carts.

> The herring man came around regular. Herring, forty for
> one shilling. That was my job as I got older, clean the harrn
> (as they were called). Kipper men also were regular callers.

Annie Marsh came around selling fish. She carried the basket
on her head. After she left George Bradney was the fish
merchant. He did well until he became a money lender. Poor
George, that was the end of fish. He got done.

Mother never had that habit, attending money lenders.
She was the opposite. She would give rather than take.

Hugh Coffee from Newburn hawked fruit and veg. He was
a rough-neck those days. He joined the army, married a
widow and then set up a drapers shop in Newburn. My mother
used to say, 'Hinny, it's not brains you want; it's plenty of
brawn and brass.' How right she was in those days, but not
now. Life has changed.

In addition to the travellers there were many smaller-scale
attempts among the people themselves to set up little shops and
some did actually flourish. What they represented, I think, was
a means of making a little money and of providing an occupation
other than housework for some of the women to do.

I remember also in Mount Pleasant and other colliery rows,
little shops. Some were huts in a garden; some were in an
outhouse in the yard. One was owned by Mrs Barry, one by
Cecil March, one by Mrs Donnison, one by Mrs Liddle. These
shops were useful for small items. They sold haberdashery,
sweets, also food. Most of these people bought stuff from the
Co-op, put on coppers, but they did get the dividend from
the Co-op for their purchases and in those days dividends
were the only means of saving money. And it was all cash.
No credit so they had nothing to lose. No shop inspectors, no
rents, no leases and no taxes. So they did pick a living from
their little shops. In fact, years later my twin brother turned
our wash-house into a shop. He also sold draperies. Our Eva
and mother made ginger beer, sold it at tuppence a bottle.
Summer Sundays they did a good trade.

I remember a fish and chip shop, the very top house of
Mount Pleasant owned by Mr Haslam. He had a good shop. He
put a big window in the outside wall. People queued up for
his fish and chips.

These activities were economically very marginal to my grandparents. As my mother says: 'It made no difference to us. We
were never any richer. We lived from day to day, week to week
and hoped for the best.' They were, however, part of the whole
setting of Mount Pleasant and represented for my grandmother
possibilities for different kinds of shopping or for stretching
her budget further.

My grandparents once took in a lodger and managed to accom-
modate him by packing my mother off up the street to stay with
her aunt Maggie. The episode indicates that they were quick
to seize opportunities to earn a bit more.

> We once had a lodger called Tommy Lynch. Dad brought him
> home for a night or two, but he liked our home so much
> mother kept him. I remember him paying thirty shillings a
> week board money – that was a big help to our family income.
> He was homeless, a bachelor and his brother put him out. I
> was young at the time. I went to sleep at my Aunt Maggie's
> up the street so I did not care what happened at home.
> Where they all slept I do not know....

Such opportunities were rare; the real income of the family was
increased not by more money, but by work in the garden or
on Tommy Lamb's farm.

Managing money was my grandmother's job and it involved
careful planning. There was no scope for impulse buying and
major items of expenditure had to be paid for, like the suits, on
an instalment plan. Budgeting was, however, an uncertain business since my grandfather's wage depended in large part on the
quality of the work place he had been allocated in the quarterly
cavelling.

> Pit work was a worry to us all and cavelling days were very
> anxious ones. Poor Dad. We knew the minute he entered what
> cavel he'd got. His face showed sadness. I think the weeks
> were counted until the next cabling [in Throckley cavelling
> was pronounced cabling].

If he had been lucky enough to get a good place where he could
make a lot of money all was well; if not, then the family budget
was cut back and my grandmother had to plan her outlays more
carefully. Mrs Thompson told me that if her father came home
with a bad cavel her mother would just go upstairs and cry. As
a child she remembers this happening often.

The quality of the cavel governed whether or not my grandfather got any pocket money. He used to give over his wages
on pay day and was given back his pocket money – rarely more
than 2s. 6d. per week. If the wage was low then he did without.
However, pipe tobacco – his baccy – was bought from the household income, and if he managed to make any money doing odd
jobs at Lamb's farm then that was considered his.

Housework itself determined the routines of my grandmother's
day but within those routines she had other matters to attend to,
particularly with regard to children and, in her case, her relatives. For bringing up children was the central motif of her life;
the housework was a mere adjunct, albeit a major one, to this
central task. It was her task, too, especially as her children
grew older, to iron out the frictions which congested living
produced and to hold the family together. As the children grew
up each took his part in domestic chores, but while they were
young in the period before the end of the First World War they
were a major worry for her, yet at the same time the source of
her deepest satisfactions.

Illness was her main worry and her main object, and defence
against it, was to keep the children well fed. Her cooking was
not something which took her away from the children; it was a
key part of bringing them up. But if she did face illness then
she felt well prepared. My mother has set out quite clearly what
my grandmother's policies were in this respect:

> When any of the family were ill, first we were put into the bed
> in the kitchen. It was always warm and we could get atten-
> tion without mother having to go upstairs.
>   The old cure for a cold was castor oil or syrup of figs. As
> she said, 'Clean out the system first.' Chest colds were
> treated with goose grease rubbed on the chest, hot drinks
> and sleep. She used to say, 'Don't worry about food; sickness
> feeds itself. Once they get the fever out of them they will
> eat. Let them sleep it off.' Then the cod liver oil and malt,
> or Scott's emulsion. I hated both. Parrishes' chemical food
> was given to us. That made us eat. She used to say, 'Feed
> a cold, starve a fever.' The sore throat remedy was awful,
> sulphur blown down our throats, then washed down with
> milk. She also used to mix equal parts of honey, olive oil and
> rum. A teaspoon of that for a cough or a black mint soaked
> in vinegar. And if any cash to spare at all, a bottle of sherry
> bought. Each day a fresh egg beat up, add 1 teaspoon of
> sugar, 1 glass of sherry and fill up with milk. Best pick-me-
> up there was.
>   With measles the curtains were drawn in case the light hurt
> our eyes. So we always got good attention. Sore throats -
> gargle with salt and water and then take a spoonful of glycerine
> and honey. Sore eyes - bathe them with boracic and warm
> milk.

The medical armoury was hopelessly inadequate but it brought a
high quality of personal care and a feeling of security in a situation where self-help was far more important than the doctor.
    These medical skills were not practised only on her own family.
My grandmother helped her neighbours and was in turn helped
by them. She attended at births and was attended in turn when
her own children were born. She helped lay out the dead, too.
Since, for her, such activities were part of a taken-for-granted
world, and in themselves of no special note, it is difficult to
discover what her involvement in them meant for her deeper
views about life and death and the significance of family life. It
always strikes me that her life was led close to the basic problems which concern all human beings; her involvement with
people, her neighbours, was direct and very close to their most
intimate concerns and not just at sad times. She, like many
others, helped at weddings, at christenings and, of course, at
funerals, and she was helped in return.
    Maintaining the links of neighbourliness and friendship was
a central theme in her working life, part of the business of

being a pitman's wife. Thinking about these things, being concerned about them was another essential part of being a housewife; without those contacts her life would have been considerably impoverished and insecure.

A WOMAN'S PLACE

The final force shaping domestic work was the prevailing social attitude about the place of women. There was no work for women in mining communities outside the home. My grandmother had worked as a shop girl but she gave that up when she married. There was, too, a clear sexual division of labour in the family. In my grandparents' home, for example, the women did nearly all the domestic work. They even dadded the pit clothes on the outside wall to get the dust and loose muck off them. The regular deliveries of house coal from the pit had to be shovelled into the coalhouse. The women even helped with that. Cooking, cleaning, looking after children, washing, shopping, etc., were all thought of as women's work.

My grandmother did not champ at the bit about housework; she got on with it and would, I am sure, have felt it strange if anyone thought it should be otherwise. And here, perhaps, is the tragedy of it. Standish Meacham (1977) has stressed the 'traditional character of working class' in the period before the First World War, its fatalism, its preoccupation with the familiar and the concrete. This description fits. There was little in the routines of my grandmother's life to contradict her whole feeling of the inevitability of it and the price to be paid for work of this kind was, as Leonore Davidoff (1974, p. 419) has perceptively noted, 'the narrowness of horizons, the closing in of the woman's world'. Just as men passed on to their sons the values of work in the pit, the women passed on to their daughters those precise qualities which made a miner's wife. The social mechanisms of this are examined in the next chapter.

There were few opportunities for any kind of release from the recurring demands of the house. For most of the time her work and her leisure were fused. As her children grew older - particularly the girls - and could help there was some easing of the burden, but never any release. As described in an earlier chapter, before the First World War when my grandfather owned a horse and trap she would go for a ride in the trap, usually up to Heddon-on-the-Wall. My grandfather would have a drink but my grandmother stayed with the children outside. Once a year she went on the Sunday school trip to the seaside. She rarely went for walks with my grandfather although he used to walk quite a lot. Indeed, in response to suggestions that she should go out she used to say that the fresh air made her 'intoxicated' and she would only trip up!

The only occasions when she could step back from the flow of inevitable chores were the great ritual occasions of Easter,

Christmas and New Year's Eve and christenings, and the great
changes in her life occurred at times of crisis as, for example,
when she took in Gordon Brown's children (see the next chapter),
or when her own children married. For most of the time she
worked.

I cannot remember my grandmother wearing anything other
than her dress and her pinny. There were occasions when she
put on her Sunday best, but for most of her life she kept to
her working clothes, a fact which states powerfully her basic
view of what her role was and how she conceived it completely
in terms of the house. Her hands were working hands, and while
she kept herself neat she was not fastidious about her appear-
ance. My mother notes in passing:

> Mam never wanted clothes as she never went far, just local
> shops. They took us to the Sunday school trip, that was a
> treat to us all. But Dad insisted on going long walks. He
> walked with us but not Mam. She was always too tired and
> she liked the house quiet just for a few hours to herself.

Within these constraints she evolved a basic attitude of acceptance; she did not complain; she found her pleasures where she
could and she did not fret about what she did not have. She
often expressed her philosophy in proverbs and she passed
these on to her children as if they distilled what she had taken
a lifetime to discover for herself. As my mother says:

> My parents always said, 'The only way to get money is to
> work for it.' 'If you want anything done do it yourself and
> you know it gets done.' 'Never put off for tomorrow what
> should be done today otherwise you get in a muddle.' So
> every day had its chores. 'Do good and good comes out of it.'
> They were both good with proverbs. I carry them through
> myself. They always said, 'Hard work never killed anybody.'
> So we all had to work.

Work, then, was the motif of her whole life, the source of
her self-respect, her access to the community around her, the
opportunity for her deepest self-expression and the quality
above all others that her husband valued. It seems to me it was
also a kind of imprisonment, a grinding necessity, but one
which could have been differently organised. But this view is
that of an outsider; for her, work was her life; she knew no
other and could not afford the luxury of dreams she could never
realise. Commenting on the position of miners' wives in his
account of working-class women before 1914, Peter Stearns (1972,
p. 108) makes exactly this point: 'because there was little sense
of alternatives there was little visible despair'. Unions and the
Labour Party offered a glimpse of a different future for the men.
What did they offer the women?

8 FAMILY LIFE: THE EARLY YEARS

The routines of housework were the scaffold of my grandmother's busy life; but what interested her ultimately were her children. In her scheme of priorities the children came first, her husband next and her close relatives not far behind. She shared this basic commitment with my grandfather although their tolerances were different. In small ways he was much firmer with the boys. He objected less than she did to the prospect that her sons might go down the pit. He was less aware than she was of the wearing effects of continuous work, although nothing in her background encouraged her to stand far enough back from her life to be very critical of it. Her sensitivity and awareness of a different way of living were, however, greater than his and stemmed, I think, from something I have been told often enough by her children. She detested the pits. She was aware of the basic limitations of life in Throckley whereas my grandfather, much more fully involved in activities outside the home where he could indulge his interests, was less critical. In fact, as I have shown, he valued much of his work, his home and his family life; the contrast between what he had achieved for himself and what he had experienced as a boy was sharp enough to convince him he was much better off. And in any case, coming from the area, he felt at home. He was well known; he had many valuable contacts, at work, on the farm, in the village. Nothing in his experience, at least up to the First World War, had prompted him to stand back from the normal flow of his life to look at it afresh.

Following the argument about domestic work and leisure I shall show in this chapter that the pattern of family life in Mount Pleasant was shaped by the routines of the pit and by the prevailing assumptions in Throckley about the respective roles of, and relationships among men, women and children. As an institution the family played a key role in reproducing those relationships, at least during the period up to the late 1920s when the village retained its character as a mining community, providing few opportunities for employment outside the pit.

Probing where I can the character of relationships inside the family, I want to illustrate further one of the central themes of this book, that private experiences of people have a social shape. The character of relationships in the family, between my grandparents themselves, between them and their children, and among the children, shaped their feelings towards one another, their sensitivities and self-images. These feelings are compounded now

into a nostalgic image of a close-knit family life. Some aspects of
the image are idealised. But the family was for them all the most
significant reference group and it gave precise definition to
certain character traits and values which, in my view, all my
relations share in a clearly recognisable way.

My grandparents' tolerance of small children is remarkable.
Four years after they moved from Heddon to Throckley, during
which time their family had increased to three children (Olive,
Jim and Eva) and my grandmother had recovered from the emotional pain of the loss of her second child, they were presented
with a problem; it illustrates both their attitudes to children
and the significance they attached to ties of family and kin. Just
before the birth of their fourth child, Bill, they had to decide
whether or not to take in and look after the children of my grandmother's half-brother. He had remarried, having met his new
wife in a pub at Benwell, a suburb of Newcastle. According to
family recollections – and these, of course, are based on the
story as it was told and retold by my grandmother – my grandmother said to him, 'Gordon, hinny, that's not the place to get
yourself a wife.' This remark is a small indication of the distance
my grandmother, like many other people in Throckley, put
between herself and people from 'doon the toon', or 'toonas' -
those who lived in the jerry-built flatlets which stretched across
Tyneside and with whom she had had to deal in her days as a
shop assistant in a second-hand store along the Scotswood Road.
Gordon's new wife clearly fitted the stereotype, for it is also said
that she used his four young children badly, her indifference
towards them turning at times to downright cruelty.

The decision to take the children into her own home was forced
on her, so it is said, by the children themselves. Feeling unloved
and neglected, and led by Lotte, the eldest, 8 years old at the
time, the four of them (i.e. Lotte, Gordon, Sadie, Jim) walked
from Benwell to Throckley, a distance of some four miles, having
sneaked out of their room under cover of darkness. They were
determined to live with my grandmother, their aunt Sally, and
they were taken in. Maggie, my grandmother's sister, took two
of them; my grandmother kept the other two and both were
brought up as her own with the full agreement of her brother
and, of course, my grandfather.

By 1905, then, she had three children of her own and, pregnant again, was boarding two others. There were seven people
living in a two-bedroomed house, a clear case of overcrowding
although quite typical for the district. From early in their
married life, therefore, their sphere of the personal and the
private was quite severely circumscribed. This affected directly
the relationship between the adults and children; there were
few secrets and no expectation that any of them should have
much right to privacy. Secretiveness, reserve and circumspec-
tion in these circumstances brought censure. Openness and being
forthright were the qualities which were valued. And both my
grandparents had a functional tolerance of noise and clamour

without which their family life would have been unbearable. They simply enjoyed the company of children. They had their own; they lived close to Maggie and to Bob and Alvina Brown so there were always lots of young relatives around, too.

But in 1912, when my grandmother was 40, tragedy struck. My grandmother became pregnant again and gave birth to twins, Louie, my mother (christened Margaret Louisa after aunt Maggie), and Jack. My grandfather celebrated the occasion by getting thoroughly drunk and on the way back from the pub he actually wore out the knees of his trousers through crawling on the ground. The birth of the twins gives a brief glimpse into the relationship between my grandparents. It was never evident to their children, despite their close congested living, that there were sexual relations between the parents – such things, in any case hidden from children, were not even discussed when, much later, the daughters themselves were approaching marriage. But my grandmother did once comment to my mother that her husband was a strong, healthy man and that if she could – meaning if it were possible at all, which by then it wasn't – she would 'still be having bairns'.

Their own relationship, like that of other couples with large families in pit houses, had to be worked through in conditions where there was very little privacy and where long hours of work left little time in the day when they could be completely by themselves. If my grandfather was on foreshift – 3.30 a.m. to 11.30 a.m. – there would have been an opportunity to be alone with his wife, assuming the children to be at school. My grandmother always got up with him and saw him out to work whatever shift he was on, and she always waited up for him whatever time he was due to come home. This was typical behaviour. Jack Lawson (1949, p. 34) calls it 'the old law of the colliery woman', 'based on grim, sad experience', which urged her to see her husband and sons out lest she never see them again. Whatever real intimacy there was between them had to be snatched from a pitiless round of commitments, clamour and growing tiredness.

When they recall their childhood my relatives do not think readily about conflicts between their parents. They did exchange harsh words on occasion; they did have their rows. But they were never serious and they were always quickly forgotten, an observation which leads to an important point about marital conflict in mining communities. Conflicts between man and wife in mining communities carry risks which are very different from those in less dangerous work conditions. Here, I think, is an example of importance of the 'row' as a social institution. Family rows were issue-specific, short in duration, in some families violent, but always quickly recovered from. The explanation for this does not lie in the personalities of the people involved; it is in the logic of working underground. To have a man go to the pit simmering with grudges and acrimony is to incur the serious risk that he might lessen his attention to his work and through that suffer accident or even death. The row clears up an issue quickly; it leaves no grudge.

Dennis, Henriques and Slaughter comment in their study of family life in Ashton that 'the development of deep and intense personal relationships of an all-round character, is highly improbable, and observations confirm the absence of any marriage corresponding to the ideals of romantic love and companionship' (1956, p. 228). They affirm even more strongly that, 'So long as the man works and gives his wife and family sufficient, and the woman uses the family's "wages" wisely and gives her husband the few things he demands, the marriage will carry on' (ibid.). And what lies behind this, they say, is that the family 'is a system of relationships torn by a major contradiction at its heart; husband and wife live separate, and in a sense, secret lives' (ibid.).

My grandfather was not a demonstratively affectionate man and nor did my grandmother expect such affection; their relationship was, however, based on a considerable respect for one another and a fundamental trust. They shared the view that the children mattered most and on this basis remained stably married into old age.

KINSHIP

One of the main attractions of Mount Pleasant for my grandmother was that her sister, Maggie, lived there. My grandfather's brother Bob, the pit blacksmith, lived there, too. They were in close contact with both relatives, and although their other brothers and sisters lived away from Throckley they maintained regular contact with them. Apart from his younger brother, George, who was a cobbler, all the Browns were connected with pits in the area. This meant, of course, that my grandfather was greatly aware, quite aside from his own experience in Throckley pit, of the imminent danger of all pit work. His younger brother Joseph, for instance, was killed in Clara Vale pit in 1914 at the age of 23. Recently married, though without children, his death was particularly cruel since his move to Clara Vale was entirely the result of falling foul of the pit management at Throckley.

My grandfather felt Joe's death badly because he was the second of his younger brothers to be killed accidentally. Harry, the tearaway, the one expelled from Heddon school, was drowned in the Tyne at Wylam trying to save a friend who could not swim. This was in 1904. Harry was 22, totally drunk and attempting, as a prank, to swim the river.

My grandfather's relationship with his own family was not really close, however, and he did not actively seek to build up close relationships between his own children and those of his brothers and sisters or, indeed, to have much to do with his own parents. From my mother's account of it there is clearly some trace of social hierarchy here, albeit of a very subtle kind:

My mother was more of a town woman, more refined, not a
rough country woman. She spoke nice and didn't swear. I
think Dad saw a lot of quality in her that his in-laws did not
have. Our house was nice and he was proud.... It did not
bother my dad's people very much ... in fact, dad's brothers
thought I was stuck up until they properly got to know me.
They were all rough and ready but very kind.

The significant relative was Maggie; the two sisters worked
closely together, they shared their children. My mother, in fact,
says she spent more time in Maggie's house than she did in her
own. Since Maggie's husband was a pit official they were slightly
better off than my grandparents and with only one daughter,
Francy, they often were able to help my grandparents financially.
They sometimes bought boots for the Brown children, for example. And since the coal they had delivered to them was of a
superior quality they often shared their coal, too, so that my
grandmother could get her oven up to higher temperatures for
some of her cooking. Their relationship was built up of a thousand sharing trivia of this kind, of daily contact and great concern for one another. The two families were, in fact, quite
indistinguishable.
This kind of neighbourliness was not unique. When Throckley
people talk of the past, neighbourliness is one of their main
themes. Not only were good neighbours inescapable, they were
necessary. A community without the means to buy in the help it
needs from the outside must meet its needs from within. Neighbourliness was what made such help possible in Throckley. It
functioned, too, to provide a framework of social contact for the
wives of Throckley (cf. Dennis et al., 1956). Albert Matthewson
captured both points when he told me about his mother:

My mother used to sick visit. I remember the Armstrong
family ... three children died of the plague (that's dysentery).
There had to be a street by street collection to bury the kid-
dies because the parents could not afford it. Living in streets
... friendly ... there's an atmosphere you don't find now.
The doors would never be locked. They used to knock through
the walls to the neighbours to see they were all right.

PARENTS AND CHILDREN: 'HAVING BAIRNS'

In Throckley before the First World War there was a very high
risk that children might die at birth. In 1905 the infant mortality
rate for the district was 191 deaths per thousand births - nearly
two out of every ten children died during the first year of life.
The average national infant mortality figure between 1901 and
1910 was 128 (Dyhouse, 1978). The birth rate and infant mortality
figures for the district for the period when my grandparents
were having their children are shown in Table 8.1.

Table 8.1 Birth rate and infant mortality rate, Throckley, 1896–1912

	Live births per 1,000 population	Deaths per 1,000 during first year
1896	35.3	123
1897	38.9	141
1898	43.8	140
1899	–	163
1900	40.0	209
1901	38.0	173
1902	38.6	170
1903	41.4	153
1904	40.6	135
1905	38.3	191
1906	35.0	111
1907	35.1	93
1908	39.6	142
1909	–	–
1910	33.1	103
1911	30.9	–
1912	30.1	135

Source: Calculated from the Annual Reports of the MOH, Newburn UDC.

Throckley, clearly like many other mining villages had special difficulties. Carole Dyhouse has suggested that a quite prevalent contemporary explanation of a high rate of infant mortality in working-class families was that of the ignorance and incompetence of working-class mothers. She says of the report of the Physical Deterioration Committee that 'its pages [were] littered with references to a new generation of women ignorant of domestic management and disinclined towards their duties in the home' (Dyhouse, 1978, p. 257).

The MOH for the district, Dr Messer, attributed the very high infant mortality rate to two sets of interacting factors, child-rearing habits and housing conditions, but before illustrating his argument it is important to take note of one of his points about the figures themselves. They are clearly very high, but the MOH added a further cautionary note in 1905:

Just think of it. 191 out of every 1,000 children born are buried within the twelve months. I have just hinted above that besides this dreadful holocaust we are liable to over-look the damaged ones.... If it gets bad food, impure air and water to drink, if it is cramped in space or cradled in dirt, how can it become anything else than a stunted crea-ture continually dependent on others for help?

As to child-rearing habits, the MOH was to point out a self-reinforcing cycle of poor diets leading to undernourishment, retarded development, and child-rearing practices in the next generation which themselves would be inadequate. In 1902 he wrote despairingly of carelessness among some mothers in the district:

> As soon as possible, after birth, the child is put onto the bottle, and whilst milk is said to form a considerable proportion of the food thus substituted for mother milk, my experience is that very little milk enters in to its composition and in its stead the most unwholesome, unnutritious, undigestible messes are resorted to. Common foods are soured milk out of dirty bottles; bread and water; arrowroot, sometimes with milk, cornflour starch and similar substances....

And of standards of cleanliness in some homes he wrote:

> Some of the houses I visited when I was enquiring into these deaths under one year were simply in an appalling condition of filth. Evil smelling, doors and windows kept carefully shut, and all manner of filth on the floor was not uncommonly met with.

He was quite clear in his own mind what caused such conditions although he looked behind the immediate causes to far deeper ones. The immediate cause was drink:

> Perhaps at the bottom of much of this vice and misery is drink, for the parents who develop this craving and spend on it what should go to the purchase of the necessities for the life of the children ... and I should be inclined to put down to this cause many of the dirty slatternly homes that one sees.

In Dr Messer's view the problems of child neglect were an artefact of ignorance.

Carole Dyhouse finds little evidence in support of this theory but, like Dr Messer, regards the behaviour of some parents as a symptom rather than the cause of the high rate of infant mortality and feels that much that was written about working-class family life in late Victorian England was 'accusatory rather than descriptive', revealing more 'about the viewpoint and values of the observer than about those being observed' (Dyhouse, 1978, p. 267).

If the business of giving birth is seen from the working-class mother's point of view a different picture emerges. In my grandmother's case it is a picture of great care being taken to protect the health of the child, and also of the birth of children being part of the life of the community as a whole, revealing much of

the character of family life and the position of women. And one
feature of that position is the matter of respectability. Dr Messer's descriptions of the poorer, more brutalised families may be
coloured somewhat by prejudice, though I doubt it, but they
do at least convey something of the range of standards of family
life against which women in Throckley could measure themselves.
For my grandmother, cleanliness was almost next to godliness,
and treating 'the bairns' properly was a central component of
her self-respect.

Prompted by such considerations my mother described to me
the preparations that were made for new babies in Mount Pleasant,
incorporating my grandmother's guide to baby care and her own
experience.

> Even in the poor days, the coming of babies was prepared at
> the first sign of pregnancy. The layette consisted of three
> day gowns, three nightgowns, three vests, three barrow
> coats, three binders (they were long pieces of white cloth or
> three inch bandages; they were wrapped around a baby's
> stomach to keep the navel intact until firm and set), three
> liberty bodices. They were to strengthen the spine and keep
> the chest warm. Cot blankets were wrapped round them –
> they were made out of old blankets or bought ones if money
> could be spared to buy them. Babies' heads were covered
> with their day gowns, that was to protect the opening of
> their heads. No light got to their eyes until they were about
> a month old.

Birth, then, had to be carefully prepared for; not to do so well
in advance would have been to incur very high costs all at once.

> Birth itself must also be seen in its social context. My grand-
> mother gave birth to all her children in her own home and all of
> them were delivered by her friend and neighbour, Mrs Allen.
> Mrs Allen also helped with funerals and weddings and pig killings,
> it being inconceivable in Mount Pleasant that any of these things
> might take place without her. My grandmother's sister, Maggie,
> was also there at the births and during each one my grandfather
> was dispatched away up the street to sit and await the result.
> Birth was a matter for the women alone. If a doctor was needed
> then he was called, but in none of my grandmother's confine-
> ments was this necessary. After the birth it was usual for the
> woman to stay in bed for ten days to recover. One aspect of this
> recovery was to wrap the mother's stomach quite tightly with
> a bed sheet in the belief that this would help the figure back into
> shape again.

CHRISTENINGS

Births are public occasions celebrated in distinctive ways in all
societies. In Throckley the period after the birth was one of
much visiting leading up to the christening of the baby and the
churching of the mother. My mother's account of Mount Pleasant
explains:

> Babies wore the long gowns and barrow coats until they got
> christened at the age of four to six weeks. Then they were
> shortened (that was dresses, all babies, girls or boys, wore
> dresses and petticoats, napkins and matinee coats and bootees,
> always no less than three each.)
>   There was always a cake baked and a bottle of something to
> drink, even if it was ginger wine. Everyone that came to visit
> a new baby got a piece of cake and a glass of wine. It was
> called 'to wet the baby's head.' And everyone was expected
> to hansel the baby with a silver coin, a threepenny piece, a
> sixpence or even a shiling if one could afford it. If one did
> not do that, that passed for bad luck to the baby.

Anyone who did not acknowledge the new baby in this way would
have been thought of as very mean indeed and not too close to
the family.
   'The christening meal was always a nice tea, all home made,
that was prepared for months in advance. In fact, it was like
Christmas preparations.' Here was the chance for celebration,
to have friends and relatives to the house, to eat, drink and
talk. But beneath the revelry there was a strict rule about the
churching of the mother and the christening itself which indicates much about the position of women, and the ways in which
they could stage their respectability.

> After the christening the mother could go out and visit but it
> was reckoned to be bad luck for a mother and baby to go into
> any house before the mother was churched and the baby
> christened. The mother was supposed to be unclean until she
> was blessed at Church.

It is clear that many women actually did feel themselves to be in
some way unclean having, as the birth of the baby unambiguously
testified, given themselves to a man. Even within marriage,
therefore, sex, given the strength of this tradition, was thought
of as being in some way dirty.

> Then again, the first three houses that were visited, the
> occupant had to give a parcel to the baby, containing three
> articles, either food, garments, mostly it was a candle, a
> box of matches, tea, sugar, etc., just traditions.
>   Also, at the christening, the Godmother that carried the
> baby had a parcel made up of eats (a sandwich, cake,

biscuits and always a silver coin put in). After leaving the
house, if it was a boy baby, the first female they met, young
or old, got the parcel. If it was a girl baby, the first male
got it. It was considered lucky to receive christening bread.

A high infant mortality rate and a high child mortality rate
are indicators of poor social conditions. But they signal, too,
some terrible, essentially private tragedies and bring in their
tow anxieties which no family with small children could be entirely
free of, especially in communities which are relatively small and
socially cohesive. My grandparents brought up their children at
a time when medical treatment for some childhood epidemics was
nothing better than an isolation policy such as closing the
school. The prospect was always a real one that their children
might not survive. A school log book entry for 17 May 1907
speaks volumes:

> This school is closed by order of the Medical Officer of Health
> for this district, on account of the Epidemic (Diphtheria and
> Whooping Cough) for a fortnight.
>   June 20th. One little girl aged 5 yrs died this morning in
> the Hospital - Annie Aager. (Throckley School Log Book)

My grandmother's eldest child, Olive, was a pupil at the school
at this time and knew Annie Aager well. Even children were
acutely aware of death. My mother recalled for me some of her
early memories of this theme:

> In Mount Pleasant all neighbours helped each other, with
> deaths, births, weddings and parties. A death was a morbid
> affair, what I can remember. The corner where the bed was
> ... the wall was draped with a white sheet. The body was
> put on the bed. That was draped in white sheets. The beds
> were black and brass, big balls on the corners and big bows
> of black ribbon were tied around the bed ends. No trestles
> those days. Fires had to be put out, wash-house fires lit and
> the wash-house used the time the body was in the house.
> Blinds drawn. It was awful. Later years, things got better,
> more modern methods.

THE WORLD OF CHILDREN

With little space and several children a constant problem was
keeping the children in order so as not to be totally over-
whelmed by their demands or distracted by their behaviour. Both
my grandparents were very indulgent to children; they had time
for them - my grandfather more so than my grandmother - and
liked their company. If, on the other hand, they felt that the
children were intruding too far into their adult world they would
promptly dispatch them into the street. 'Get away and divvent

cock your lugs here' was a common remark if any of them tried
to listen in to an adult conversation.

The world of the children was, in Mount Pleasant, quite
separate from that of the adults. It was, however, just as constrained by the resources of the village as the position of the
family was by the pit, and was just as powerfully shaped by
prevailing attitudes about the respective roles of men and women.
It was, then, a constructed world and its nuances reveal much of
my grandparents' attitude to their children and how they interpreted their own roles as parents.

For the children themselves the world was bounded by the
street and the neighbours, the street being their main playground. My mother once again explains:

> As children we got very little entertainment, just what we
> made ourselves. In Mount Pleasant, plenty of kids, two long
> streets to play in, street lamps on either side, so around
> these lamps we played, that is in the dark nights until about
> seven o'clock. If wet we gathered in our wash-houses with a
> big fire on. We played games (Guess What), very simple
> entertainment, dominoes, card games and always under the
> eyes of our parents.

Their play changed according to the seasons of the year and,
given the complete lack of any special facilities for children's
play in the village, their ingenuity had to express itself in
meagre surroundings:

> In the summer nights we played in the burn, a gutter behind
> the Lyric Picture Hall. It was walled up on each side. We
> used to pile sticks and stones to dam the water. That was
> our bathing pool. Each girl had their own pool. We used to
> wash the stones, sandstone them and very proud we were.

There is a connection here, clearly, between the girls' play
and the housework routines of the home, a subtle form of anticipation which acted out in play their fate as housewives. One
of the favourite pastimes for the girls was to take babies out in
their prams. My mother says they used to knock on doors where
they knew there were babies and ask to take them out, anticipating in play a common enough task for mothers. The sexual
division of labour, clearly, was in part reproduced in the play
of children.

Unlike boys, girls did not expect to work when they grew up.
Apart from casual domestic work and some posts as shop assistants there was no work for women in the village. Just how
difficult things were is revealed in a minute of the Co-operative
Store Committee in 1895. It reads:

> Annie Scott to be appointed apprentice in the drapers'
> shop:

Conditions:- 12 months without wages
　　　　　　　to learn millinery
　　　　　　　to serve in shop whenever required
　　　　　　　One month's trial on either side. Throckley
　　　　　　　District Co-operative Society, Minutes,
　　　　　　　30 September 1895, Tyne-Wear Archive 1062/1.

That same month sixty applications were received for the post of
grocery assistant. For the girls, the key role was that of wife
and mother.

The absence of proper play facilities often led the children
into dangerous areas. My mother related that above the burn
there was a 'big cundy', a large open pipe.

I don't know where the water came from but it was fresh and
clean, never stagnant. The cundy must have been miles long
and we used to crawl up it with a candle lit. That happened
often until my father heard about it. A good telling off I got.
He said, 'Bairn, don't go up there. That water is from the
pit. Rats will come down.' I think he said 'rats' to scare me;
never again did I go up the cundy.

There may have been a very special reason behind my grandfather's warning. Throckley miners had suspected for a long time
a close connection between pit rats and Weil's disease, otherwise
known as rat jaundice. Local doctors confirmed the connection
in the 1930s.

Children regularly went for walks. 'Summer nights we roamed
the Dene, paddled in the burn, walks up the river side.' During
the winter nights, 'we played at the lamp - ball, skipping ropes,
boys, leapfrog, Jack shine the lamp, block, knocky-nine-door,
but our parents were always popping out to see if we were O.K.'.

For the boys the pattern was slightly different from that of the
girls. They were destined for work, a fact reflected in my
grandfather's attitude towards them and in the kind of qualities
which their play valued. For them childhood ended when they
started work, and if they were to work in the pit it ended at
14 years of age.

My grandfather made certain that his sons contributed to the
work of the family, especially in the garden. And Jim, his eldest
son, remembers as a boy in 1913 having to look after all the
gardens while his father lay in bed recovering from an accident.
'I used to go to the garden with my father lying in bed. He'd
cry out: "How many tetties have you set today? Divvent do over
much."'

By his own example my grandfather sought to instil in the boys
the supreme value of work. He himself, as I have already
explained, worked, when he could, part-time at Lamb's farm.
He was always busy. Work for him was something to be got on
with without complaint. Bill and Jim discussed this with me, and
Bill began: 'We never lost any time. Work was first with father.

1 Heddon colliery, *c.* 1908

2 Throckley Front Street, 1904

3 Mount Pleasant, Throckley

4 Throckley co-op staff, 1907

5 Throckley colliery, 1908

6 Throckley colliery winding gear, 1930

7 Shaft bottom, late nineteenth century

8 James Brown (far right) and marras, *c.* 1914

9 James Brown and leeks

10 The Brown family in 1905

11 Throckley miners digging coal, 1926

12 Grandparents, 1931

13 Louie, Eva, grandparents and Gloria, 1934

14 Grandparents' golden wedding anniversary

15 Grandparents outside Mount Pleasant, 1952

16 Final photograph: James Brown, aged 92 years, at Mount Pleasant

If there was no time for work there was no time for pleasure. It was as simple as that.' Jim added, 'Aye, there was no idle time.' He used to tell his boys that if they had not got a job before they were 14 then they would have to go down the pit.

This particular attitude was reflected in his views on their education. He made sure they went to school but he did not expect much to come of their schooling. Nothing in his experience had led him to value school, and as the boys got older he tolerated their staying away from school to do a job for someone and even allowed Bill to leave school prematurely to secure a job at the Throckley laundry.

Albert Matthewson's comments on his own youth confirm what my uncles have told me about theirs. The boys played billiards in the store hall. In the summer they played pitch and toss at the 'hoyin' skeul' by the muck heap on the back lonnen. This always brought them into trouble with the police and they went to elaborate lengths to create an advance warning system to avoid detection.

Football, pigeons, fishing and simply hanging around corner ends were all part of the boys' lives. It mattered greatly whether particular boys could hold their own in fights; to be thought a 'cissy' was a profound tragedy. The qualities which were valued were toughness, daring and an ability to model adults in an authentic way. Being able to take a hammering like a man and to do battle fairly were important qualities. In that short boyhood it was a crime to tell tales, to blab to parents or teachers. And to make the transition from the boyhood world to the rituals of courtship was a chancy business. To do it too soon would risk ridicule or being spied on by a stealthy group of mates eager to announce their presence at a critical moment. To do it too late brought other kinds of sanctions: accusations which questioned a lad's basic manhood. Shyness, reserve and sensitivity were not qualities which brought much respect although they could be tolerated. As in the case of Heddon, the Throckley coal company could rely on the habits of the boys themselves to build up in their employees those qualities of toughness and masculinity without which work underground cannot be undertaken.

CHAPEL LIFE AND CHILDREN

Chapel life was important for the Brown children although they were not formally members; they considered themselves to be Church of England people. They did, however, go to chapel Sunday schools and were on occasion members of the Independent Order of Good Templars which, in Throckley, was supported strongly by both chapels. The chapels were important institutions to most children in Throckley and had been so since they were first built. From 1891 onwards, the year in which the Primitive Methodist chapel was built at the top end of Mount Pleasant, there were two chapel circuits active in the area and

both made a special appeal to children through their Sunday
schools and support of either the Band of Hope or the Good
Templars; the Primitives actively proselytised through the
'Christian Endeavour'. The Sunday schools were very well
attended, there being something of a tradition of Sunday school
attendance: as early as 1877 over 150 children on a Sunday
afternoon and sometimes more than 200 (Throckley Wesleyan
Sunday School Minutes). In 1900 the Primitive Methodist chapel
regularly attracted 155 scholars. Both circuits organised 'socials'
for children and young people.

The values which, in general, Methodist Sunday schools sought
to uphold have been discussed by Robert Moore (1974). They
included sobriety, self-respect, obedience to those in authority,
hard work leading to individual fulfilment, truthfulness, forbearance and, of course, the love of God and a practical application of the faith. And what lies behind the teaching of such
values is a vision of society which is disciplined, caring, sober,
chaste and honest in which men and women live their lives in an
unostentatious way in the sight of God, contributing to their
communities.

How such values were actually transmitted is partly revealed
in my mother's account of her involvement with the Good Templars. The organisation she says, was 'simple and quiet':

> It was always on a Tuesday night, held in a classroom in
> Throckley school from 6 to 9 p.m. It was run by a man called
> Mr. Graham and a Mr. Cook It was a mixed meeting, boys
> and girls aged 10 to 14.

Incidentally, the fact that it was run by two men is not surprising. Most Sunday school teachers in Throckley were male. As
early as 1877, from a total of fifty-three named teachers in the
Wesleyan chapel, forty-five were men. In 1893 in the Primitive
chapel eight of the nine teachers were male.

> The main part was a service like a chapel or church, more
> chapel, as Mr. Graham run the chapel at the top of Mount
> Pleasant.
>   We were selected as officers. I was the chaplain. I was a
> good clear reader so I read out the prayers. We all wore
> regalias around our shoulders. That showed we had a posi-
> tion. One person arranged games, another conducted to our
> hymn singing, another arranged entertainments, concerts,
> socials. A small collection was taken to keep up expenses. If
> we made a little party our parents baked cakes and made a
> few sandwiches. Some nights we just had a good sing song,
> all sacred songs. Mr Graham gave us talks.

But at the heart of this religious activity was a central social
message and a core ritual:

> We all signed a pledge when we joined. This is the pledge:
> We promise by divine assistance to abstain from all intoxicat-
> ing liquors, tobacco, gambling and all bad language.
>   Mr Graham brainwashed us that to do all the things I have
> mentioned caused poverty, trouble and made broken homes,
> money spent unwisely. Bad language was disgusting and very
> bad for children to hear. Also stealing and telling lies was
> very bad. I remember Mr Graham had a short arm and he held
> a gavel in his hand and he always shouted, 'Silence' and hit a
> wood block with his gavel. At once he got silence. We all
> jumped to attention. We thought he was the most wonderful
> man in the world.

The only tangible evidence of her few years in the Good Templars
that my mother retains is a battered old lead medal which she
often refers to jokingly as the only formal honour she ever got
in her childhood.

The chapels were not unique in having an extensive social
presence in the village. The church, too, ran its Sunday schools
and socials, the Mothers' Union and even a sewing class which
children were encouraged to attend. These were, then, important
institutions in the lives of children and, as my mother stresses,
most children in the village joined in the activities of the chapel.
To do so was, she told me, 'a kind of duty; just like going to
school'. And they all went in their best clothes, changing back
into their 'glad rags' when the services were finished.

The amount of care lavished on children and the extent to
which their lives were regulated, particularly by Sunday schools,
are things which the German miner, Dückershoff, commented
on particularly. The fact that children attended day school
impressed him immensely since it gave women more time to themselves, a contrast to Germany. And of the children themselves
he had this to say:

> Children are nicely dressed, for the workman is properly
> proud of his bairns. On a Sunday they are not to be distin-
> guished from those of the well-to-do. A great deal is done
> for them. Almost all attend Sunday school, and during the
> summer joint excursions are arranged by these schools,
> generally on Saturday afternoons. (Dückershoff, 1899, p. 54)

Dückershoff is sensing here what was involved locally in
'doing right by the bairns'. And I think, too, he is raising
implicitly another aspect of the way in which miners and their
wives could stage publicly their claims to the respect of others.
Adults could, in a sense, project themselves through their
children. To have well turned out children who went to church
or chapel in polished shoes and carefully ironed dresses was
an unambiguous sign that within the privacy of the family
there were high standards.

IMAGES AND EXPECTATIONS

When my relatives talk about their early childhood they evoke a
picture of a contented, happy family. They do not recall their
childhood as being strictly regulated. Nor do they regard themselves as being different from anybody else in Throckley. I
have pressed them firmly on this, always with the same response.
The Browns, they insist, were just an ordinary family; they did
not feel themselves to be different. They remember being busy,
the girls especially. By the time Olive was 13 she baked the
bread for the family. Both Olive and Eva helped a great deal with
the twins and the housework. These older girls were drafted
quickly into the work routines of the home. They recall, too,
Olive playing the piano, and quiet evenings when they simply
sat indoors, my grandmother busy knitting and the old man
sitting with his clay pipe. Olive's piano recital always ended
with the Blaze Away March and my grandfather saying, 'We've
had it now; that's the finish.'
Despite the image, however, there were divisions. Eva and
Bill bickered constantly. Jim fell easily into the role of eldest
son and spent a lot of time with his father. My mother spent as
much time as she could at aunt Maggie's to avoid the congestion
of home. The boys and the girls did not play together. The
twins did and they all felt protective towards Jack whom they
regarded as a delicate child. The older girls had separate groups
of friends and so did the boys. Olive and Bill spent more time
in the house than the others and they all quarrelled among themselves over whose friends could visit.
There was, as I have shown, a sexual division of labour in the
family, and the respective male and female roles in the community
appeared clearly in the relationships between my grandparents
and their children.

THE BOTTOM DRAWER

The boys were expected to work in the gardens and to 'hold their
own' against others. The girls were expected to prepare for
marriage. Here is the importance of the so-called 'bottom drawer',
the method of collecting and saving for married life those articles
of household linen or decoration which would be difficult to
provide in the early stages of setting up a home. The bottom
drawer is a key part of the girls' memories of their childhood
and youth, and from it they themselves acquired an image of
their own married lives.
Like most other women in the village, my grandmother kept a
bottom drawer for her daughters although not for her sons. In
her case, the drawer was quite literally the lower drawer of the
wardrobe, but that is not important; it is the concept of the
bottom drawer that matters. From childhood onwards she collected items of household linen like pillow cases, sheets, table-

cloths and embroidered antimacassars for chairs which her daughters would need when they set up their own homes.

As the girls grew older they themselves added to the collection
of items, lavishing great care on embroidery and crocheted pieces
such as the edging to a linen tablecloth. Carefully ironed, these
items were stored with either mothballs or lavender bags, to be
taken out only when the drawer needed tidying or, in aimless
days, when the girls took them out to anticipate in holding them
the days when they themselves would be married. The bottom
drawer was a focal point in the relationship between my grandmother and her daughters, offering an opportunity for marriage
to be discussed in a practical way, and for her to give advice
and consolidate in her daughters' minds those subtle values of
decency and respectability which could be symbolised in neatly
embroidered covers and clean, well-ironed sheets. It had, too, a
clear economic rationale. The bottom drawer was built up of
necessities and luxuries, but always of the smaller things which
it would take the daughter some time to assemble after she had
married.

I wondered whether discussions about the bottom drawer ever
led to questions about sex. My mother assures me they did not.
'We learned about that the hard way.' My grandmother was
clearly prepared to advance her opinion. She told her daughters
often enough if they heard of anyone falling pregnant that 'it
just happens to the good ones', and, invariably added, 'She's
not the first and she'll not be the last!' But she never discussed
the matter in any detail with her daughters.

Generally, however, their lives were lived with an intense
immediacy, responsive to the varied routines of the family and
the village, totally unburdened by any expectations that they
themselves should aim for a life different from that of their
parents. As I show in subsequent chapters, this changed, and
was part of the subtle metamorphosis of Throckley itself from a
mining community into an urban village. Under the impact of
war, the long-term decline of the pits, political change and industrial diversification, Throckley acquired a new character. In the
period being discussed, however, it was still very much a mining
village with its values and relationships intact.

One final point: the picture of their early family life my relatives now reconstruct has at its centre a powerfully nostalgic
image of their parents. The family, it seems, revolved around
the parents; the children were devoted to them, especially the
girls. This 'home-centredness' had quite tangible consequences,
as I shall show. For the moment it is sufficient to note the image;
it is central to their sense of the past and it gave meaning to the
pattern of kin relationships which developed as they all grew up.
It still feeds their sense of what has been lost as the price of
social change.

9 THE FIRST WORLD WAR

At the outbreak of war in August 1914 my grandfather was 42
years of age, too old for compulsory military service and with
too many dependents to contemplate volunteering for the army.
Of his six children the eldest was only 14 and the twins had
just turned 2 years of age. He was at the peak of his powers and
fully occupied with his work and his gardens. Since his sons
helped him a bit and his eldest daughters helped out in the
house he could see the way forward to being a lot better off
than he had ever been before.

Not only did he not want the war, he thought it was unneces-
sary. Like many other people he had expected war and had,
through his union, been led from as early as 1905 to oppose any
such developments and to condemn militarism. Perhaps, then,
he agreed with William Straker's July circular in which a stance
of 'absolute neutrality' was demanded:

> We have no quarrel with any of the great nations of the world
> that would justify the shedding of a single drop of human
> blood, and yet while I am writing, the cloud hangs black and
> threatening over our land, as it does over all Europe, and
> may burst at any moment in a deluge of blood, unequalled in
> the world's history.

Straker attributed the international position to capitalist interests
and corrupt diplomacy and, calling for neutrality and for a conference of the Triple Alliance and the Miners' International Conference to condemn war, he warned, 'Just as suddenly as war
has broken out in Europe, revolution may break out in Great
Britain.'

It did not take long for the tensions and anxieties of July to
evaporate in the jingoistic certainties of August, for many miners
to join the army and for Straker to find himself somewhat isolated
and criticised for his stance against the war and for mixing his
trade unionism with politics. In his August circular, war having
been declared, Straker made an appeal for national unity and

> the necessity of all differences of opinion being dropped; of
> civic and party strife being ended; of industrial disputes
> being settled; of the terms 'employer and employed', 'rich
> and poor' being forgotten; and only that we were Britons
> being remembered.

These sentiments were to evaporate as the war dragged on, but
for the early part of it they were sufficiently strong to command
wide support. Local newspapers reported extensively on recruiting campaigns, and recruitment in the mining districts, was,
indeed, brisk. The 'Report of H.M. Inspector of Mines' for 1914
(p. 32) notes:

> Since the British Empire became involved in war, the miners
> of the Northern counties have responded to the Call to Arms
> in a manner which has kindled very pardonable local pride;
> and the large numbers which have joined the Forces and the
> enthusiasm which they have displayed in the work of training
> will be remembered with very keen satisfaction for many a
> year to come.

Some 52,000 men left the Northern Division to become soldiers,
of whom 13,000 were from Northumberland. A high proportion
were married men.

Two factors specific to the coal industry in the north helped
military recruitment among miners. The mines inspector noted
one of them which affected married men particularly:

> The way has been made clear for them by the munificence
> of the colliery proprietors who undertook to make such pro-
> vision for those of their own workmen left behind as would
> ensure them being able to live in comfort. (ibid., p. 33)

They noted one firm from which so many men had been enlisted
that it was costing £70,000 per annum to maintain the families
left behind.

Throckley coal company was no exception. On 13 August 1914
at their monthly meeting they made arrangements to remunerate
the wives of miners who left for service at the rate of seven
shillings per week and one shilling per week for each child under
16. In the first wave of recruitment from Throckley eighty-one
men, of whom thirty-seven were married, left the pit (Throckley
Coal Company Records, NRO 407). The inspector of mines summed
up his report on recruitment with an observation which has great
significance for understanding the long-term decline in labour
productivity in the pits which the war did little to reverse. 'One
must admit', he wrote 'that, as a rule, it was the flower of the
various classes of labour which offered its services to the country'
(ibid., p. 33).

The second factor helping the recruitment of miners was the
virtual collapse of the export trade in coal (Redmayne, 1923,
p. 9). In his September circular, William Straker showed that,
among other effects, this market collapse caused a great deal of
short-time working. In 1913 in Northumberland the average
number of days per week worked was 5.42; in August 1914 it was
2.72 (NMA Minutes, 1914, September Monthly Circular, NRO
759/68). In some pits unemployment was high as well, and these

factors contributed directly to encourage men to leave the pits, especially when at that point there was no expectation of a protracted war of attrition in the mud of Flanders.

Gordon Brown, the lad my grandfather had helped to bring up, was quick to join up, having been a member of the Northumberland Fusilier Territorials. Jack and Jim Breckons, my grandfather's nephews whose younger brother William had been killed at Heddon pit, also joined up. Many of those who went were good friends of my grandfather, but he thought of Gordon as a son. Uncle Bill remembers Gordon leaving. He had received the call to go to Newburn barracks very early. He left aunt Maggie's house in his uniform, carrying his kit bag and saying, as he kissed the women and children goodbye and shook hands with the men, 'Don't worry, this lot won't last long.' 'And that', said Bill, 'was the last we saw of him.' Eva recalls he took with him from the top of the piano a photograph of his sister Sadie, saying that if he was 'knocked out' they would find the photograph and think it was his 'pretty girlfriend'. He died in France at Ypres in 1915, seven days after leaving home. Aunt Maggie got the telegram informing them of his death on the same day that she received a letter from him saying he was well. Aunt Eva says she remembers being in the house with her mother and aunt Maggie when the news came. My grandmother, she remembers, sat 'rocking her chair with grief and wringing her hands'. They heard later that he was not killed outright. Badly wounded, he was moved on a stretcher to the sanctuary of the church near the front. Moments later the church was obliterated by heavy artillery fire. 'This news', said my uncle Jim, 'nearly killed Maggie.'

The boy from two doors away was killed in action. Mrs Allen's two sons, Hughy and Billy, were killed. And so it went on. They tried, so my uncle Jim says, 'to mek the best on it'.

The First World War was a turning point for my grandfather just as it was for the village, the pit and the whole mining industry. I want to show in this chapter that the war reinforced demands among the miners for the nationalisation of the pits, that it stripped politics of its respectable aristocratic veils and spawned, both nationally and locally, a Labour leadership bent on social reform and rooted in a determined and radical working class confident of its political and industrial strength. In Throckley the most enduring and positive effect of it was to create a very determined local Labour Party, which, while pressing for control of the urban district council, managed to persuade the council to build houses on a massive scale. I shall show, too, however, that the confidence which the strength of their wartime position gave the miners, and which resulted in what they took to be the great victory of the Sankey Commission with its recommendation that the pits should be nationalised, was short-lived. For in the longer term the war did little to strengthen the bargaining power of miners and much to weaken it. And for many ordinary miners, my grandfather included, the broad-based

optimism for the future to which the ending of hostilities gave
vent quickly evaporated into a cynical despair that none of it
had been worthwhile.

The first six months of the war were lived out almost in an
atmosphere of business as usual. Indeed, the local newspapers
for this period still devoted more space to local items and sport-
ing events than they did to war reporting. The September leek
shows were unaffected by hostilities as was the tail end of the
cricket season. Yet much was already changing. The letters
columns of the 'Blaydon Courier' were filled almost immediately
by complaints from readers that some local traders were putting
up prices, a practice which was unanimously condemned as
unpatriotic. The harvest festival at Throckley school was that
year celebrated with the aim of sending its proceeds, together
with satchels and bandages, to the Armstrong hospital for
wounded soldiers. The newspapers were already publishing
horrific accounts of the experiences of the British Expeditionary
Forces and early in November began publishing a roll of honour
listing the deaths of local heroes. In Throckley and elsewhere
in the urban district, street collections were organised by the
local war relief fund and by October £376 had been raised.

Fund raising went on throughout the war. The Throckley
colliery band held regular socials and concerts in the store hall
in aid of the Throckley soldiers' relief fund, and there were
occasional visits of male voice choirs as in July 1917 when Wallsend male voice choir sang to 'a large audience' in the store hall
to raise cash for the relief fund. Groups of women were organised
by Mrs W.F. Stephenson, the wife of one of the coal company
directors, to knit and sew for the comfort of the troops, and my
grandmother regularly contributed wool socks.

A small but significant blow to my grandfather came with the
requisitioning of his horse by the army. He willingly gave it up,
but although not having to look after it eased his commitments
it made some of his work harder, particularly feeding the pigs,
and it took away some of his pleasures, like the Sunday ride out
with the family in the trap. With shortages, requisitions, changes
in working hours, not to mention the terrible news from the
front, the war closed in on people quickly and completely. ·
Throckley school was used in the Easter of 1916 to billet soldiers
from the Tyneside Scottish battalion, and when they left they
were given tea at the Wesleyan chapel. Aunt Eva says that the
soldiers were no bother in Throckley. They marched and drilled
and the children imitated them. And, of course, 'the lasses
loved it: they lapped it up'. Similar arrangements elsewhere in
the district and throughout the war brought many soldiers to
the village, and those returning on leave added to their number.

Of course, it was those returning who could confirm with
graphic immediacy the full horror of the front. Albert Matthewson
told me that his brother George came back, and before he was
allowed into the house he had to be deloused. From being initially
unwilling, like thousands of others who returned, to talk about

it or even to make contact with his old pals - such soldiers were
pained to be asked the obvious question, 'when are you going
back?' - he gradually let them all know the awfulness of it. And
soldiers' tales did nothing to promote confidence in the quality of
leadership traditionally claimed by the upper-class military personnel of the army. Bob Cowan, an old Throckley miner who
fought in France and returned to the pit afterwards to become a
union official, often told me in long conversations about the war
that the officer class were beneath contempt. In Bob's case this
view carried over from the men themselves to the whole social
order they symbolised. And it was not an uncommon reaction; it
had wider political credence in the way Lloyd George replaced
Asquith over the military mismanagement of the Dardanelles
(see Marwick, 1968).

My grandfather could never find words strong enough to sum
up just how he viewed the First World War. 'Bye hinny,' he
once said to me, with that momentary expression on his face
which clearly showed him back in time recalling it all, 'we divvent want ti gan through all that again....' And while much
remained the same for him after the war, it none the less had a
profound effect on him. He was not a demonstrative person and
he kept his feelings well under control; he went to the street
parties in Mount Pleasant which celebrated the Armistice and
he attended, as an onlooker, the Remembrance Day parades
which followed annually after the war, but he never released
his grip on the sickening feeling of disgust and cynicism which
displays of shallow patriotism induced on him. He never developed unpatriotic attitudes; that residual pride in his country
which his education had instilled in him was never abandoned.
And he certainly did not succumb to the war-time myths of the
evil Hun. If the war did anything at all to the national stereo-
types, he held it reinforced his view (and in this he may well
have been influenced by returning soldiers) that the French
were a spineless lot and totally unreliable as allies.

What disturbed him was warmongering; he used to say, for
example, that Churchill was 'nowt but a warmonger', and it was
one of his main criticisms of the Tory Party as a whole that its
public patriotism concealed a lust for war. Aunt Eva is convinced
that the First World War explains my grandfather's intense
dislike of Tory politicians. How deep and long-lasting that dislike
was emerged when, as a very old man, he watched Churchill's
funeral on television. 'Instead of a state funeral', he said, 'they
should have tipped the bugger over the bridge.' It was the war,
I think, which, by affecting the patterns of daily life, revealed
which social groups in the order of society were really vital -
the direct producers of wealth - and reinforced the work of the
unions and the Labour Party of the period up to 1914 in strip-
ping the respectable veneer off Edwardian society, exposing a
mechanism which could no longer command anyone's respect. For
my grandfather the experience was a profound one; through
war he saw more clearly than his earlier and somewhat uncon-

scious attachment to the Liberal Party had ever allowed that progress for the likes of him presupposed a much more fundamental
change in the order of things than he had previously imagined.
What he was not too sure about was how the needed changes
could come.

The question of prices was one of the factors which fed his
disenchantment. As already mentioned, in the early days of the
war there was a great deal of resentment about price increases
in the shops giving rise to charges of profiteering (see Marwick,
1968). And as the war went on and the basic cost of living
increased these charges found concrete political expression. In
May 1915, for example, in his monthly circular William (Bill)
Straker connected rising prices to rising profits and pointed
out:

> Shipping profits have been enormous; profits in mining,
> although badly hit in some districts at the beginning of the
> war, are now following close on the heels of shipping. Arma-
> ment manufacturers scarcely know how to dispose of their
> profits, and wholesale dealers in foodstuffs are in the same
> position and all the while there are hundreds of thousands of
> the poor, especially in our large cities, lacking the neces-
> saries of life because of enhanced prices. (Northumberland
> Miners' Association Minutes, NRO 759/68)

By 1916 this concern, at least in some quarters, had developed
into a full-blown criticism of the management of the war. In the
'Blaydon Courier' of 13 February 1916, H. Blakey of Winlaton,
a regular correspondent and miners' sympathiser, had the
following to say in response to proposals to increase the length
of the miner's working day:

> I am willing that sacrifices should be made, but I am not
> willing that all the sacrifices should be made by the indus-
> trial classes, and particularly by the mining portion of them,
> while others, as those, for instance, who are gambling with
> people's stomachs in wheat markets make fortunes out of the
> sacrifices. Let the strain of war be equally distributed all
> round and if it is to be felt anywhere more severely, let it
> fall on those who are best able to bear it, and by those whose
> sordid interests all through the world's history have been the
> cause of war.

On 11 November 1917 Newburn Urban District Council discussed
the price of foodstuffs with Councillor Browell, the miners'
checkweighman from Blucher pit and leading Labour activist in
the area, moving a resolution 'viewing with alarm the ever-
increasing cost of the people's food' and demanding action
immediately to regulate prices and food distribution 'so as to
minimise the possibilities of exploitation'. The food distribution
question had another aspect which created local difficulties in

Throckley. In 1916, under government pressure, the co-operative store terminated all credit on payments for foodstuffs thereby removing one means of overcoming temporary hardship for
some families.

It was, indeed, the prices question in the north-east which
qualified the general conclusion of the Commission of Inquiry
into Industrial Unrest that, on the whole, the working classes
in the area had taken up a 'sane and patriotic view' of the war.
What they said was this:

> Joined to the sense of actual hardship, there is undoubtedly
> a deep-seated conviction in the minds of the working classes
> that the prices of food have risen not only through scarcity
> but as a result of manipulation of prices by unscrupulous
> producers and traders who, it is alleged, owing to lack of
> courageous action on the part of the Government, have suc-
> ceeded in making fabulous profits at the expense of the con-
> sumer. (Commission of Inquiry into Industrial Unrest, 1917,
> p. 2)

Set alongside the daily reports from the front line, the evidence
on price rises and profiteering caused great bitterness which
developed into a questioning of the whole political management
of the war, and from that to the organisation of political life
itself.

The issues of conscription for military service which became
acute in 1916 (in May of that year universal conscription was
introduced following the failure of the earlier voluntary schemes),
taxation and the gut politics of injustice can be mentioned here
for they illustrate the connection between the immediate experience of war and changes in political consciousness. They are all
related to one another. As early as 1915 Bill Straker had bitterly
attacked Lord Northcliffe (and his press) for 'creating distrust,
fear and hatred between the nations in order to produce war',
and in 1916, on the question of conscription, Straker contrasted
the risks which the wealthy faced in the war with those faced
by the common soldier. Should not wealth, he asked, 'share the
same fate instead of being only borrowed to be fully paid back
with a substantial increase in the shape of interest. The slain
men can never be given back to those from whom they are taken'.
Straker's campaign against conscription was a hopeless one; in
August 1917 he had to inform the lodges that the coal controller
had agreed that 21,000 young miners should be recruited for
the War Office, of which Northumberland's share was 955, the
number to be drawn from each pit depending upon the number
of men employed. Fifteen young men from Throckley were con-
scripted under this arrangement.

The conscription question was linked by the union to the
proposals to increase income taxes in 1917. There was a meeting
organised by the Throckley and District Miners' Federation in
the co-operative store hall in October to campaign explicitly

'against the imposition of income tax on workmen' ('Blaydon Courier', 13 October 1917). The meeting was chaired by Dick Browell and was addressed by the indefatigable Mr Straker. Two resolutions were submitted and adopted: 'That as the measure of all taxation ought to be the ability to pay, we protest against the taxation of workmen's wages under £15 per annum.' And:

> in the face of the conscription of men, we assert that all war expenditure ought to be met by the conscription of excessive wealth, and ask the Government to pass into law at the earliest possible date a 'Conscription of Wealth Bill.'

Here, then, is some evidence of the escalating anger of miners against the conduct of the war which led to a sharp consciousness of class divisions. And this consciousness, fuelled by a growing anger about questions of politics, was reinforced for the miners by their experience in the pits.

Right from the very beginning miners were encouraged to see themselves in a new light: as playing a vital role in the whole war effort. In October 1914 there was a national appeal by the War Savings Committee to persuade people to save coal and light on the grounds that for 'the purposes of the war coal is an asset of supreme value' ('Blaydon Courier', 14 October 1914). Within such a climate it was particularly galling, therefore, for men to have to face unemployment and pit closures. A letter from a 'patriot' to the 'Newcastle Evening Chronicle' on 7 August 1914 articulates this theme very well and indicates, too, something of the criticism which was levelled at mine owners and which grew in the course of the war to demands for the outright nationalisation of the pits:

> No class in this country has made greater fortunes during the past few years than the mine owners of the North, yet, at the first sign of trouble and additional expense, they must needs shut down the mines and pay off their men. Why cannot they face the situation patriotically, keep the pits open, even if only partly worked, and stock supplies for a time? I think the action of the Northern mine owners is scandalous and displays an entire lack of patriotism at a critical period.

Because of the firm local market Throckley colliery did not suffer like some of the coastal pits although there were redundancies and some short-time working. The figures for Throckley Isabella are shown in Table 9.1. Employment at the Throckley Maria colliery remained a little more stable, but for the company as a whole the number of employees dropped from 2,290 in 1913 to 1,813 in 1915. What did remain among the men was a determination to maintain and even improve the level of union membership. In 1916 the Executive of the Northumberland Miners discussed a request from the Throckley lodge 'for leave to come out on strike

in order to force non-union members into the Association', and
while the Executive could not support them they none the less
decided to exert pressure on the Throckley management to secure
a full union membership at the pit.

Table 9.1   Numbers employed and union membership, Throckley
Isabella colliery, 1913-19

	Nos employed	% of employees in Union
1913	921	63.1
1914	772	74.8
1915	677	75.7
1916	716	72.6
1917	724	71.8
1918	543	96.1
1919	656	80.1

Source: From Northumberland Coal Owners' Statistical Returns
and Annual Accounts of the Northumberland Miners'
Association.

Work in the pit did not change much during the war except
that it got harder and was done by men with a higher average
age than before the war; throughout the industry output per
man fell heavily (Court, 1951).

Like many other small undertakings of this period the management at Throckley did not respond to the war-time demand for
coal by adding more machinery to their capital (see Technical
Advisory Committee, Report, 1945, Cmnd 6610). Indeed, the
only technical innovation the company introduced was to build
up the sides of the coal tubs by another 4 inches so they could
carry more coal. Despite the growing use of mechanical coal
cutters and conveyor haulage the Throckley coal company used
traditional mining methods well into the 1930s. The general
change in wages, particularly for coal hewers, was upward with
war bonuses, which the Miners' Federation secured at 1s. 6d.
a day in 1917, being paid to all miners. But, as Rowe pointed
out, the increase in wages was not as great as the increase in
the cost of living during the war period (Rowe, 1923, p. 90).

Until the last year of the war, relationships of authority and
power in the pits themselves did not change substantially,
although many miners did build up a resolve that at the end of
hostilities the mines would never return to the old style of pre-
war management. In 1915, for instance, it was still possible for
the Throckley management to act severely for the most trivial of
offences. The Executive minutes of the NMA show that George
Mitford was sacked in 1915 'for using abusive language to the
undermanager'. He was subsequently reinstated, but the incident
is a revealing one. And in 1916 the coal owners still retained

the right to evict widows from colliery houses and to withhold
compensation payments to the wives of miners until they actually
left the property. In addition to the normal controls on miners
to keep to their work diligently there was, from 1916 onwards,
considerable pressure from the government, filtered through
local managements and, in 1918, enforced by pit production
committees, to eliminate absenteeism, the threat being that the
persistently absent would be called up for service in the army.
What had changed in the pits, however, were the legal, political
and ideological conditions of coal production. By the end of the
war the government, largely in response to labour unrest in
South Wales and the need to maintain coal supplies, had virtually
taken over the running of the whole industry (see Kirby, 1977).
On the miners' part the pre-war resolve to win the wholesale
nationalisation of the industry had become an urgent political
demand. This was clearly the policy of the Miners' Federation of
Great Britain and it was widely supported. In 1917 the Northumberland Miners' council meeting of November passed the following resolution from the Throckley branch:

> That we continue to press for the nationalisation of mines
> and royalties and the abolition of way leaves, so as to relieve
> our industry from an unjust impost and to secure to the na-
> tion the value of its own natural resources.

And the gradual extension of state power throughout the war
prompted Straker to give his members some general political
advice:

> We are evidently becoming more Socialistic - with a difference.
> The difference lies in the fact that the state does not belong
> to the people, as it would under socialism; but that the people
> belong to the state, and as the state is made up of only a part
> of the people the danger is that one part will be enslaved by
> the other part. (NMA, Minutes, monthly circular, March 1917,
> 759/68)

The following month Straker responded to the events in Russia:
'In the name of Freedom we greet the revolution in Russia.'
Later in the year, worried about how the burden of war debt
might be paid off, he forewarned his members that such debts
might be erased by reducing wages and creating inflation. He
raised the spectre of 'the inevitable seething discontent which
will take possession of the working classes, when, a few years
after the end of the war, severe depression sets in and hunger
is gnawing at their vitals'. This same mood was maintained by
the union throughout 1918, and in December of that year they
widely circulated the membership with the advice of the Miners'
Federation of Great Britain: 'Remember. The Coalition is a coalition of landlords and capitalists - a coalition of Wealth against
Labour - a coalition which will do its best to prevent Labour

realising its aim.'

The news of the Armistice, although awaited, came with the
force of a surprise on the afternoon of 11 November. Many
people from Throckley went off to Newcastle by tram (they
could do so now, the tram-lines having been completed in 1916)
to join in the celebrations. In the days and weeks that followed
there were street parties, but there were also memorial services
in the chapels and the church, and the Christmas of that year
was particularly solemn. During this period, too, a war memorial
committee was formed with Mrs W.E. Stephenson in the chair. In
the spring of 1919, with a parade of troops led by the colliery
brass band, the war memorial was officially unveiled with fifty-
one names inscribed in gold letters.

Amid the hectic uncertainty of the immediate post-war world,
but not in the least persuaded that the world would ever settle
down again, my grandfather picked up again some of his earlier
plans. Jim, his eldest son, was just about to leave school, and
under pressure from my grandfather work was sought, not in
the pit, but at Spencer's steelworks in Newburn. Olive was 18
still living at home but working in domestic service again (having, like thousands of others, worked in a munitions factory -
at Blaydon) and courting a lad from Crawcrook over the river.
Eva, having left school without a job, earned a little of her keep
doing housework as a 'day girl' for various neighbours, then
succeeded in getting a job in the store. Both girls, then, were
earning some money and Olive was in sight of being married.
Since the twins were just 6 years old life went on much as
before for my grandmother, and my grandfather had still to go
to the pit.

But one small fact highlights just how much had changed, as it
were, beneath the surface, and this was my grandfather's
absolute refusal to allow his eldest sons, both of whom were of
an age to do so by the end of the war, to join the local Boy
Scouts. He told them it was just a training in militarism and that
it didn't matter that it was connected with the church. Perhaps
he was here again taking his cue from Bill Straker who, at the
beginning of the war, attacked the National Service League and
the church itself for using patriotism to discipline men. Straker
said,

> The snares of this movement are laid for the feet of the child-
> ren, and many parents, deceived by the glamour of it, allow
> their lads to join Boy Scouts and Church Lads' Brigades
> utterly unaware of the real purpose the military party in this
> country have in view.

He wrote that these organisations sought 'to exploit the worker's
sense of patriotism and appeal to his fighting instincts; to discipline him to absolute obedience by making a soldier of him'.
Finally, and very bitterly, he said: 'In the churches where the
Prince of Peace is supposed to be worshipped they conduct their

blasphemous parades.' And he ended by quoting Shelley when he wrote against the idea of the natural depravity of men, which these youth movements sought to overcome through their discipline:

Nature? - No!
Kings, priests, and statesmen blast the human flower
Even in its tender bud: their influence darts
Like subtle poison through the bloodless veins
of desolate society.

The mood, then, was one of disenchantment and cynicism and the feeling was strong that the post-war world would have to be very different. A Newburn district councillor supported this evidence. Speaking of the need to remove the tolls from the Newburn bridge across the Tyne, he warned his colleagues, most of them, at that time, being Liberal: 'The public conscience of the district has been quickened to a sense of justice' ('Blaydon Courier', 7 July 1917).

The feeling that things must change was widespread. Jack Lawson put it well in 1945 when he wrote somewhat cynically about the unredeemed promise of the first war that those returning would never go back 'to the old evil conditions under which the masses had existed'.

We did at least learn that the gentleness which makes men great is neither the handmaid of wealth nor social standing. If the war did nothing else, it pricked the bubble - and the gentleman whom we call the worker knows it too. They may have returned to the old conditions, but the old superstition of a superior people who are entitled to a superior life has gone forever. (Lawson, 1949, p. 146)

The ideological atmosphere to which Jack Lawson alludes was, of course, not without its contradictions. In the general election which followed the war the coalition government, with Lloyd George as prime minister, was able to continue capitalising on end-of-war euphoria with promises of a land 'fit for heroes' and backed by a Parliament in which the majority of MPs (338) were Conservative but in which the Labour Party, with sixty members, was the largest opposition group (Miliband, 1973, p. 64).

In the Wansbeck division, the constituency for Throckley, the miners' candidate, Ebby Edwards, was defeated by a coalition Liberal, R. Mason. Among miners, however, there was a very strong feeling of resentment about the results of this election; they believed that it had been called too soon and that working people had been misled by promises which could not be kept. In some cases discontent took a distinctly marxist tone. George Harvey of Wardley, the firebrand checkweightman of Follonsby pit, demanded much more than social reform; he wanted revolution. In a pamphlet issued at the end of the war, advertised

with the injuction 'Order Them Now and Equip Yourself for the
War after the War', Harvey insisted that 'our very lives are now
in the grip of the capitalist machine' and that it was time to
fight for possession, control, and freedom. The Executive of the
Northumberland Miners would have no truck with Harvey, declining to buy any of his pamphlets. But although Harvey was a
unique maverick his ideas were not so far away from those of
some ordinary miners, especially those schooled before the war
by the Independent Labour Party (ILP). In Throckley, for
instance, a lodge resolution of 1919 adopted a very radical tone:

> It was unanimously agreed at a mass meeting of Throckley,
> Maria, Blucher and Heddon workmen that all pits be idle on
> Monday, July 21st, as protest of the following resolution:
> 1. Protest against the Government using British soldiers
>   for the destruction of Russian Democratic Revolution.
> 2. Protest against the continuation of conscription.
> 3. Protest against the imposition of six shillings per ton
>   being raised on the price of coal.
> (Steam Collieries Defence Assocation, Minute Book 16, NRO)

Danny Dawson was one of the new generation of local Labour
leaders, schooled in the ILP, active in the union and a founder
member of the Newburn and District Local Labour Party. In passing this pro-Soviet resolution the Throckley men were clearly
echoing the views of their union leadership who believed, as
Straker put it in his January 1920 circular, that international
capitalists were 'tormenting' 'poor bleeding Russia' to restore
'the old rotten, traitorous, tyrannical regime of the Tzars, in
order that they may have the opportunity of exploiting the
natural wealth of the country and this in the name of liberty'.
    Preparing for the war after the war suggested to some miners,
too, that they must improve their education. A resolution from
a Blucher miner, Tom McKay, to the committee of the Throckley
co-operative store in February 1919 brings this out:

> That we, the Throckley Co-operative Society Limited, affiliate
> with the Central Labour College with the view of sending two
> students every year; also starting Educational Class in the
> district on 'Marxian Economies' [sic!]. (Throckley District
> Co-operative Society, Balance Sheet: 1919. To be deposited
> in Tyne-Wear Archive)

And to emphasise finally that the prospect of bringing about the
new social order was for many Labour activists one which
threatened revolutionary politics of the sort which the end of
hostilities had brought in Germany and which the Russian Revolution had itself signalled, Bill Straker can be once again quoted.
In his January 1920 circular criticising Winston Churchill and
Lloyd George for their attacks on the Labour Party, he wrote:

I have said that 'the class war will not commence with the
workers'. Neither will revolution; but it may be forced upon
the workers when the workers come into political place and
power, and attempt to pass laws of equity and justice.

This, then, was the ideological atmosphere in which post-war
reconstruction began for many miners in the coalfield, and in the
Newburn-Throckley area this quickened consciousness of conflict was sharpened by a small but significant group of ILP
members - Danny Dawson, Dick Browell, George Curwen, and
a few others less well-known - who, as co-operators, Methodists,
union officials and, in Browell's case, councillors, were well
placed to form the local Labour Party and to focus their attention on what for them was the critical problem of the area -
housing.
It would require a full study in itself to chart in detail the
rise of this group to a position of very significant leadership
in the area, and in their broad outline the results such a study
would produce are familiar enough already. The Methodist chapels
gave them their social conscience and debating skills; the Labour
Party and the union sharpened their committee skills and political
wit, and because they were well-known and sympathetically
respected they could fill positions of leadership in the union and
in the community. What they set out to do, therefore, is something which says much, not just about themselves, but about
the men who elected them and *their* aspirations, *their* sense of
what was politically important. Such men embodied the political
aspirations of the whole community, not always effectively, not
always unanimously, but their political work cannot be seen
outside the preoccupations of the community as a whole in
response to its own problems of work, housing and general living conditions.
In this sense, the Throckley leadership which the first war
spawned must be seen as the expression of working-class political
culture in the area, of what E.P. Thompson (1968, p. 13) has
called (admittedly writing of a much earlier period) the 'agency
of working people, the degree to which they contributed by
conscious efforts to the making of history'. The essence of
working-class culture is not, as Harold Entwhistle has recently
and forcibly pointed out, to be found in recreations and pastimes,
not in 'that amiable sociability which is popularly supposed to
characterise working-class culture', but in solidarity and political
consciousness (Entwhistle, 1978, p. 120).
Recognising, as E.P. Thompson did, that the aspirations of
workers are 'valid in terms of their own experience' (1968, p. 13),
it is hardly surprising that living in housing controlled by the
coal company would lead pitmen, though not inevitably, to seize
on council housing as the real symbol of their emancipation, and
to see control of the local authority as their principal means of
securing it.
Dick Browell, elected chairman of the Newburn and District

Local Labour Party on 27 July 1918, set out its aims:

> To further the interests of Labour in the Constituency and
> unify Labour generally for the emancipation of themselves,
> by having full representation in Parliament and Municipal
> Urban and Rural Councils and Boards of Guardians. (Newburn
> and District Local Labour Party, Minutes, 1918-27, NRO
> 527/B/1)

Dick Browell, then, a councillor since 1905, was chairman,
Danny Dawson was secretary and Bill Graham treasurer. They
were affiliated to the Wansbeck Divisional Labour Party and
through that supported the League of Nations, extending into
the post-war world their vaguely Liberal internationalism and
their belief in international parliaments which Browell, certainly,
had come to believe in during the war.

10 THE GENERAL STRIKE AND THE MINERS' LOCK-OUT, 1926

Historians differ about the way in which the First World War changed British society. Philip Abrams (1963) suggested a few years ago that the spate of social reform which followed the war was a 'failure' from which only propertied middle-class women derived any real benefit. Arthur Marwick, on the other hand, insists that the war 'pushed the state in the direction of collectivist social legislation undertaken on behalf of the lower sections of the community' (Marwick, 1970, p. 120). In my view, both are actually correct, the difference between them being that Abrams is writing about the short-term measurable effects of the war and Marwick the long-term transformation of political attitudes and expectations. In the mining industry and in mining communities these shifts in attitude expressed themselves as an insistent demand that the state should play a central role in the running of the industry and that living standards of miners should be increased. Both expectations, fuelled by an early post-war optimism, were to be dashed on the rocks of protracted industrial struggle which culminated in the defeat of the General Strike and the slow agony of 1926.

In Throckley the early post-war optimism focused very largely on the prospects of a housing programme to combat overcrowding in the district. In the pits hopes were high that nationalisation of the industry was a real possibility. The irony was, of course, that the overall position of the industry had been seriously weakened by the war. An ageing labour force, a drop in output per man-shift, the accelerating replacement of coal by oil and the failure of government reorganisation plans all contributed to that weakening (Kirby, 1977). These weaknesses were pitilessly exposed by the collapse of the post-war boom in April 1920 and the fall in coal prices. This collapse destroyed any hope that the proposals of the Sankey Commission would ever be accepted. The dawn broke, then, on a period of conflict which in 1921 led to the miners gaining, through the so called Datum Line Strike, an increase in wages; but by the end of 1926 they had lost nearly all they had gained since August 1914.

Some warning of troubles to come came in 1920 in a dispute over the sacking of Dick Browell, the Blucher checkweightman, an event the Throckley group of unions interpreted as victimisation of a Labour and union activist. A court, in a case in which Browell was accused of fraud, found in his favour and there were great celebrations in the district. The main meeting to celebrate Browell's success was held in the store hall and opened with the hymn which begins:

> These things shall be; a loftier race
> Than e'er the world hath known shall rise
> With flame of freedom in their souls
> And light of knowledge in their eyes.

I asked one of my old respondents whether this hymn did really
capture the spirit of the meeting, which had been addressed
by the union leader, Straker. Memory failed him but he did say,
'if they'd been singing the Red Flag I might have believed it.'
My grandfather was absolutely convinced that Browell was being
victimised and referred to the coal company at this time as
'bloody scallywags'. The point I wish to stress, however, is that
in this early period there was considerable hope of a better
future.

FAMILY LIFE

The early optimism of the industry had its parallels in the family
life of my grandparents. Olive was married in 1920 to a former
soldier and miner, and went to live at Crawcrook. Aunt Eva
recalls that my grandfather was practically drunk for a week.
Olive's husband caused them some worries for 'in drink' he could
be violent, but they did quite like him otherwise. Eva, their
second daughter, had left school and was employed in the store
millinery department and she was courting, too. Jim, the eldest
son, had his job in Spencer's steelworks and this pleased my
grandmother greatly. The opening of the Throckley Welfare
recreation grounds under agreements made in the 1920 Coal
Mines Act had increased the play facilities available to the twins.
There was much, indeed, to convince them that their lot was
improving.

But as the children grew so did their demands for autonomy
and space, and both accentuated the hopeless overcrowding of
the house. My mother noted that if the girls wanted to wash they
had to carry water upstairs to the bedroom. If they wanted a
bath they had to wait until everybody had gone to bed. Privacy
was precious but more or less impossible to attain and this led
to a great deal of bickering among them, although the old man
remained aloof from it.

Inevitably, however, the fortunes of the industry affected
family life. A four-month strike in 1921 over wage reductions
was auspicious. The collapse of the Triple Alliance left the
miners to fight wage reductions themselves and they lost. 'Black
Friday', the day on which the railwaymen and the transport
workers withdrew their support, became a potent symbol of
betrayal in the mining industry. And in the collective memory
of the Northumberland men, 'the miners had been starved into
submission' (Davison, 1973, p. 44). A soup kitchen had been

opened in Throckley - although my grandfather would not allow
any of his family near it - and the store paid its dividends early.
Store sales dropped from £230,563 in 1920 to £160,244 in 1921
and average credit per member went up from £1 to £5 (Throckley
District Co-operative Society, Quarterly Balance Sheets). The
medical officer of health noted for 1921:

> Unemployment on an unprecedented scale had been the lot of
> the majority of the inhabitants for the greater part of the
> year. Malnutrition must have been the lot of many for many
> months on end, and although the condition is not apparent to
> the casual observer, it is nonetheless widespread. (MOH,
> Newburn UDC, Annual Report, 1921)

For the Browns the problems were not so severe; the gardens
kept them fed. But 1921 is remembered as an awful year. Alvina,
my grandfather's sister, died that year leaving a young family
of six children. And Bill, who had left school to get a job in
Throckley laundry, was made redundant, and forced to work in
the pit. But work at the pit was scarce and there was a lot of
short-time working. For my grandmother the most taxing part of
1921 and the years which followed was the sheer unpredictability
of her income. Bill told me: 'Those days my mother did not know
from week to week what she had coming in.' And she was very
worried, too, about her daughters. Eva married in 1923. Her
husband was a miner and from the beginning the marriage
floundered; Eva had to move back to her parents with her own
daughter, Olive. My grandfather was ill with pleurisy in 1922
and 1923 and had three spells off work. He had a small income
from Heddon Club, the Friendly Society into which he contributed all his working life, but it was not sufficient. In 1925 he
suffered another accident in the pit, again through a fall of
stone, in which the tendons of his left hand were severed. He
was off work for just one week and could only hold a pick by
bending his fingers round the shaft with his good hand. Till
the end of his life he could not hold a fork properly. In 1924 he
had damaged his toes in a fall of stone, but he continued to
work, hobbling badly, driven by a grim need not to loose time,
despite his sons who tried to prevent him from going to the pit.
For lurking behind him, as he saw it - realistically, given his
age - was unemployment and the detested 'Means Test'. Unemployment in the district was running at about 25 per cent and the
Throckley pits were shedding labour.
When my relatives are prompted to think back to these years
their recollections are not entirely dominated by industrial and
political troubles. As a family they were certainly on the defensive, but they still had a good time occasionally. Grandchildren
came to Mount Pleasant regularly. Olive's daughter, Francy,
was born in 1921; Eva's daughter, Olive, in 1923. And every
weekend there was a sing-song with Olive playing the piano. The
family piano stool still contains all the music they ever possessed

and, like fossils in rock strata, those tattered sheets tell their own tale of the decline of a music hall tradition giving way to the music of the stage and dance hall. They range from the tear-jerking working-class songs like The Holy City and Danny Boy to what Richard Hoggart (1957, p. 157) once called 'the cheeky, finger-to-the-nose-ain't-life-jolly songs' such as Horsey Keep Your Tail Up. And outside the home my grandfather found what pleasures he could in the club and in the garden, although by this time gardening had taken on the mantle of necessity.

Yet there was precious little in their lives in the way of escape from the economic realities they faced; they could not stray too far from the daily need to muddle through. The whole meaning of muddling through was transformed, however, in 1926 when my grandfather, like thousands of others, was pitched inexorably and against his will into a very bitter struggle which sharply politicised his daily routines.

THE GENERAL STRIKE AND MINERS' LOCK-OUT

The year 1926 has a special place in my account of my grand-father. It was for him a turning point. The events of that summer pitched him into an awareness of his position which his busy routines had previously stifled. His consciousness of class was sharpened; his understanding of industrial action was enriched. The lock-out brought into sharp definition the parti-cular strengths and some of the weaknesses of the village, testing severely the quality of community and family life.

The solidarity, ingenuity, tolerance and strength of Throckley people were brought to the forefront of the struggle. However, the collective strength of the mining community, born of such qualities, was shown in the end not to be sufficient. For men of my grandfather's generation the experience was a profound one; it resulted, I think, in a feeling that little could be gained through industrial action. For younger men it bred both political cynicism and a determination that labour needed stronger political organisations.

In the short-run, the overwhelming mood was one of defeat and despair and great bitterness. The bitterness was directed not just at Baldwin or Churchill or, indeed, the government as a whole and their allies the coal owners. It existed, too, among the men themselves; it turned against blacklegs; it turned against those who prolonged the lock-out; it turned against those who in one way or another had made money from the trou-bles; it turned against the police. The bitterness is still recalled and is still a potent force in Throckley among older people.

For miners as a whole it helped to confirm their self-definition as a maligned, exploited group. And that is an element of their collective biography which they have never forgotten. All the institutions which the miners had built up - the union, the co-operative store, the social clubs, the Labour Party and,

in some ways more importantly, the networks of neighbourly
help - were mobilised to conflict. So, too, were the resources
of families themselves.

The events which led up to the miners' lock-out and the
General Strike have been partly described above, and extensively
set out in a number of recent histories (Renshaw, 1975; Noel,
1976; Farman, 1972; Phillips, 1976). The underlying causes of
the General Strike and the struggles in the coal industry are
clearly related to post-war attempts to stabilise the economy and
restore sterling to something like its pre-war parity with the
dollar. The costs of this financial orthodoxy were to be met
inevitably by lowering wages - and wages in the coal industry,
for very specific reasons connected with the way that industry
was organised, were to bear the largest part of the reduction
(see Renshaw, 1975).

The miners were, in this sense, engineered into a conflict
situation; they could not do less than protect the level of wages
they had achieved after the war. The logic of the situation for
the coal owners was different; in order to protect their investment in the face of a falling export market and a government
policy aimed at maintaining sterling at a high level, they were
forced into demanding both district agreements and wage reductions.

The result was a classic conflict situation where the parties
involved would look for allies to strengthen the sanctions they
could apply and to realise their interests and legitimate their
position (see Rex, 1961, for a definition of these terms). In the
case of the mine owners, allies were sought in the government.
The miners looked to the Triple Alliance and the TUC. The
conflict groups which emerged were the government and the
organised labour movement, both poised to mobilise all their
resources for the struggle but with neither group fully in control of the situation.

The logic of the coal owner's situation had to be traced to the
post-war profitability of the industry and its organisation. The
period up to 1924 had been one of rising prosperity and profits.
But after the resumption of production in the Ruhr, British coal
was exposed, as Kirby (1977, p. 68) put it, 'to the realities of
the long-term market situation'.

The issue which had to be resolved was who was to bear the
costs. The coal owners wanted to transfer the costs to wages,
proposing on 1 July 1925, for example, wage cuts of as much as
48 per cent for Northumberland and Durham. The miners were
determined that there should be neither wage reductions nor
extensions to the working day, and that the costs of the fall-off
in trade should be borne by profits and the government.

The negotiating strength of the Miners' Federation was
strengthened by a promise from the TUC General Council in
July 1925 that they would support the miners in their struggle
to defend living conditions. And on 25 July the trade union
movement pitched the conflict to a new level with the announce-

ment that the railwaymen, transport workers and seamen would embargo the movement of coal if the lock-out notices which had been served on the miners in June were not withdrawn.

At this point a reluctant Baldwin was forced to act. Having galvanised the opposition of the TUC even further by announcing that 'all workers of this Country have got to take reductions in wages to help put industry on its feet' (quoted by Kirby, 1977, p. 73), Baldwin was forced, against cabinet opposition, to agree to a continuation of the coal subsidy at not less than £10 million for a period of nine months, during which time yet another commission of inquiry could examine the position of the mining industry. These decisions of 31 July were thought of by the miners as a victory and the day became known as 'Red Friday .

Whether the continuation of the subsidy and the effective postponement of conflict in the coal industry were the outcome of delaying tactics pressed on Baldwin by right-wing cabinet members like Churchill and Joynson-Hicks and Lord Birkenhead to give them time to build up a more effective emergency organisation (see Renshaw, 1975), or a genuine attempt by Baldwin to play a conciliatory role (see, for example, McDonald, 1975), is not something I can resolve. It bought time, in fact, for both parties. The government used it to strengthen the Office of Maintenance and Supplies and other emergency measures; the miners used it to mobilise both their own membership and public opinion.

The royal commission which the government set up to inquire into the coal industry was chaired by Sir Herbert Samuel, a senior Liberal politician and former governor of Palestine. It included Sir William Beveridge, Kenneth Lee and Sir Herbert Lawrence, men who were, respectively, an academic, a businessman and a banker. In his August 1925 circular, Bill Straker described them dismissively as 'men who have scarcely any knowledge of what they have to enquire into'. While the commission sat industrial attitudes on both sides hardened and, yet more ominously, the fortunes of the industry continued to deteriorate. Had there been no subsidy, the cabinet was informed in April, 90 per cent of the tonnage in Durham and Wales and 100 per cent in Northumberland would have been raised at a loss (Kirby, 1977, p. 77).

This same point was made by Bill Straker in his February 1926 circular and it underlines heavily that for the miners the conflict in the industry necessarily involved the state. 'Without a subsidy', Straker explained, 'many coalmines in many districts of Great Britain would be closed down and thousands more miners, with their families, thrown into a state of semi-starvation. This applies especially to coal exporting districts such as Northumberland'.

The degree to which the coal dispute was politicised is indicated in a speech to miners given by Arthur Cook at Lanchester in County Durham on 8 January 1926. Cook, the firebrand secretary

of the Miners' Federation of Great Britain (MFGB), of whom
Arthur Horner once said that when he spoke he 'spoke *for* the
miner and not *to* him' (quoted in Bellamy and Saville, 1976,
p 40), told the meeting that he expected a political crisis in the
course of the year. The whole speech suggested the inevitability
of severe conflict:

> Well might the British public ask: 'Is peace possible in the
> mining industry?' As representative of the men I declare
> emphatically: 'Yes' but only under one condition. The price
> of peace must be, in a few words 'Safety and economic
> security.' By economic security we mean a wage based upon
> the cost of living at least not less than 1914. Whatever else
> may happen during this year as in 1924 the miners' motto
> will be 'No retreat, no compromise in the hours and wages'
> and I shall advise the miners not to even meet the employers
> to discuss such questions. Whatever the decision of the com-
> mission might be the miners would not consider for one
> moment the abolition of national agreements, increase of
> hours, or reduction of wages. They could not compromise
> on these three points. Therefore it is quite clear that con-
> flict seems inevitable....
> Any attempt that will be made politically to suspend the
> seven hours day will be met with united resistance from the
> whole labour movement. ('Blaydon Courier', 9 January 1926)

The negotiating stance which Cook distilled into the famous
words 'Not a minute on the day, not a penny off the pay' was
something the miners clung to well into the summer and which,
in defeat at the end of the year, they had to concede completely.
  While the three major parties to the conflict manoeuvred, the
Samuel Commission produced a report. The report was equivocal
on a number of questions, but it made it clear that the industry
needed to reduce its costs through lowering wages and that
nationalisation was not a solution to the structural problems of
the industry. The proposals to municipalise the coal trade, to
nationalise mineral royalties and to encourage amalgamations of
collieries were not sufficiently central to either coal owners'
interests or miners' demands to avoid the indifference with which
the Samuel Report as a whole was received.
  Negotiations between the three main parties in the period from
March to the end of April were fruitless. The coal owners,
dominated by the export districts, insisted on wage reductions;
the miners would not accept this without further suggestions
being made about reorganisation in the industry. Behind the
rhetoric of the different negotiating positions, the meetings
between the government and the TUC, and the frenzied reporting of the press, the conflict inherent in the logic of a declining industry, a determined leadership among the miners and
TUC support of the miners' living standards was taking definite
shape. The lock-out of miners began on 30 April. Three days

later, triggered ostensibly by the refusal of the workers on the 'Daily Mail' to print an editorial 'for King and Country', the government broke off negotiations with the TUC, turning the industrial struggle into a constitutional one to face the challenge of what had been a key element of the TUC's negotiating position, the threat of a general strike.

THE GENERAL STRIKE IN THROCKLEY

The Throckley pits closed on 30 April. There was no excitement about that; rather there was a weary sense of the inevitability of a protracted dispute which nobody really wanted. Nobody at that stage knew, of course, what kind of conflict they were entering into. Some members of the local Labour Party sensed that they needed to be well organised. In March, for example, they passed a resolution committing themselves to gain control of all the co-operative societies in the area 'with a view to the closer co-ordination of the Political Labour Party and the Distributive Co-operative movement' (Newburn and District Local Labour Party, Minutes, NRO 527/B/1). They were quite clear, too, that they did not wish to have any contact with the Liberal Party in the struggles ahead. And on 17 March they agreed to initiate moves to set up a trades council for the area. No trades council was actually set up until after the General Strike had started and then it was in response, not to the Labour Party, but to the Executive Committee of the miners' union. News of the General Strike, three days after the posting of lock-out notices, changed the mood of people, especially the younger men. For a brief moment it seemed to many of them that they might win.

The first couple of days of the lock-out, the Saturday and Sunday, were just like any other weekend. On Monday the General Strike began and by Tuesday its effects were obvious. There were no trams, hardly any cars on the roads and everywhere was quiet. The ponies had been brought up from the pit and the children went along to the fields to see them and give them treats of food scraps.

By Wednesday the mood was changing, especially among some of the younger men. There was a quickening of temper and a sharpening of the need to be organised and to do something. This was happening throughout the coalfield as several calls were made to form councils of action. A famous call came from the Spen and District Trades and Labour Council in their 'Strike Bulletin' (no. 1) on 4 May, urging miners to 'Form Councils of Action: All Behind the Miners'. It insisted, under the guiding hands of Robin Page-Arnot of the Communist Party and Will Lawther of the Durham Miners' Association, that 'the General Strike is ALREADY A SUCCESS. Do not believe the lies put out by the Capitalist press' (see Garside, 1971, p. 194; also General Strike Pamphlets, Gateshead Public Library). It ended optimistically with the injunction: 'Be of good courage, and victory is ours.'

On the same day Northumberland and Durham General Council
Joint Strike Committee was set up. Page-Arnot was active here,
too, ensuring some strike organisation to cover the area under
the control of the Government Civil Commissioner (see Flynn,
1926). The first meeting of this group, in which most of the
larger unions of the area were represented, ended on a rather
paradoxical note. The meeting, writes C.R. Flynn,

> terminated with the first hint of the difficulties of a general
> strike in the shape of a complaint that the Miners' Clubs
> faced with a drink shortage were sending in motors for beer
> whilst Transport Workers were out on strike.

Within a few days the committee was in almost permanent session.

In Throckley the most obvious feature of the strike was the
large number of miners, particularly younger ones, picketing
transport on the main west road. The local mineworkers' federation, the group composed of lodge officials from each of the
Throckley pits, tried to persuade people to keep well within the
law and avoid trouble, believing that disturbances would reduce
the moral force of the miners' case in the eyes of the general
public.

The leadership of the Throckley lodges had by this time been
given over to men who were far less committed in a political sense
than those who had been in control throughout the First World
War and the early 1920s. Danny Dawson left the pit in 1924 to
become the Labour Party agent for the Wansbeck division and
Dick Browell, while remaining in local politics, was forced through
unemployment to seek work outside the pits and resign his union
duties in April 1926. Jimmy Mitford and Bill Avis were well
respected but they were by no means radical in the union work.

However, on Thursday, 6 May, trouble did break out in
Throckley and it brought the strike right home to my grandfather. For the first few days he had not bothered much about
the strike, spending his time in his gardens. But on the Thursday uncle Jim was arrested for picketing and told he would be
summonsed. The disturbances of 6 May were on the main Hexham
road by the Throckley school. In response to calls from the
Joint Strike Committee and urged on by dispatch riders distri-
buting 'Strike Bulletins', a large crowd of Throckley pitmen -
police evidence gives the figure as between 400 and 500 - began
picketing traffic. The picketing started early in the morning,
at about 6 a.m. and went on until midday when the police were
able to disperse the group and make arrests. The picketing was
in some respects spontaneous; it was not well organised and the
police clearly thought it violent. In evidence to the magistrate
on 21 May, they argued that stones were thrown at lorries,
windows broken on buses and threats issued that if drivers
attempted to cross the lines their vehicles would be tipped over.

Uncle Bill told me that he had walked up to the road just to

see what was going on, and when he saw the crowds he came
hurriedly back home to bring his brother, Jim. My grandfather
would not go picketing; he said he did not want to get himself
involved with troublemakers. Jim and Bill then went back up
the road and by pure bad luck - he was simply among the crowd
the police swooped on - Jim was arrested along with thirty-two
other men. The police were obviously looking for leaders. It is
clear from instructions issued before the strike, during it and
long into the miners' lock-out that the police were on the look-
out for those described as 'disaffected' and 'communist agitators'
(Chief Constable of Northumberland, File: Official Circulars re
Emergency, NRO NC/1/20 1926). One man, Oliver Akenside, was
singled out in court. The police files describe him as 'disaffected'
and he was sentenced to one month in prison without the option
of a fine.

My grandparents were very worried about these events.
Miners, they knew, were being imprisoned and they feared a
heavy fine. Aunt Maggie eased things a bit, saying she would
help Jim pay a fine if that was what he got, but this did not
allay their worries. What the affair did, however, was to make
my grandfather aware of just how determined the police were in
the strike; the arrest of his son brought an immediate reality to
the many reports coming back to Throckley from elsewhere in
the coalfield; of harassment, arrests and the prevention of
meetings.

On Friday, 7 May, in response to a miners' Executive Committee request, a Newburn and District Trade Union Council of
Action was formed with John Carr of North Walbottle in the chair
and with 'Henna' Brown and Danny Dawson representing Throckley. Their first meeting was preoccupied with the safety of men
in the pit, picketing and beer supplies. They resolved 'that we
approach the officials of Social Clubs in the area and ask them
to refrain from ordering any further supplies of beer etc during
the present stoppage' (Dawson Papers, NRO 527/B/12).

On 9 May they resolved to have a picketing committee in every
village. But this determination to picket collapsed in the face of
police provocation. On 11 May they heard a report that

> pickets had been in operation at all strategic points round
> the district; but the leaders had brought in many reports
> that the police had been interfering with them, and informing
> them that the Law did not allow pickets to stop Motor Traffic.
> In some cases the police had taken up a threatening attitude.
>   Resolved that all picketing be suspended for the present....
> (Dawson Papers NRO 527/B/12)

It was difficult for ordinary people to know quite what was
happening. Two of the Newcastle newspapers managed to appear
during the dispute, the 'Newcastle Journal' and the 'Newcastle
Chronicle'; both were against strike and supported the government and the 'Chronicle' was sometimes referred to in strike

bulletins such as the TUC's 'British Worker'. The Chopwell-
based 'Northern Light' and the Newcastle Trades Council's
'Workers' Chronicle' were by no means widely available (see
Mason, 1970). Some got through to Throckley. Most men, though,
relied on word of mouth for their information and the air was
thick with rumour. The rhetoric of the strike bulletins was that
of class warfare, although this was less true of the TUC's
'British Worker'. The 'Workers' Chronicle' of 19 May might have
urged: 'Workers! On with the fight. Demand the Resignation of
the Forger's Government. Up with a Worker's Government!' But
the TUC was emphatic. On 11 May the 'British Worker' insisted:

> The workers must not be misled by Mr. Baldwin's renewed
> attempt last night to represent the present strike as a political
> issue. The trades unions are fighting for one thing, and one
> thing only, to protect the miners' standard of life. (No. 1,
> 11 May 1926, Gateshead Public Library)

Because of a lack of precise information and the uniqueness of
the situation many ordinary men were in the dark about what was
really happening. Reports of violence, arrests, communist
agitators, police harassment, blacklegging volunteers and, in
Northumberland, the celebrated derailment of the Flying Scots-
man at Cramlington, fuelled forebodings of trouble.
    The view in the Brown family was that the situation was out of
control and that they could not see an end to it. And my grandfather, it seems, was very pessimistic at this stage. He believed
the miners could win but he did not trust the TUC or, for that
matter, as I shall explain later, Arthur Cook. But he did not
want to settle on the owners' terms; he was clear on that issue.
    The view of the General Council Joint Strike Committee by
Friday was a more optimistic one. 'On Friday', wrote C.R. Flynn,

> the success of the general strike appeared completely assured.
> It was clear to everyone that the O.M.S. organisation was
> unable to cope with the task imposed upon it. The attitude of
> the population was favourable to the strikers and unfavour-
> able to the government. There were no disturbances, the
> trades unionists maintained almost perfect discipline. There
> was no change from the ordinary except for the quietness in
> the streets and the absence of traffic. (Flynn, 1926)

The quietness of the streets for my grandfather signalled
uncertainty; trouble and worry and Jim's impending court case
hung like a dark cloud over everything and there was no money
coming in. Jack and Louie, the twins, were just about to leave
school and neither had any prospect of work. He felt very much
on the defensive and his poor opinion of politicians drove him
even more to look to his own needs and not to rely on unions
and political action.
    Given that the housekeeping burden fell on my grandmother's

shoulders this was a difficult time for her, particularly since she felt she had to help Eva and keep the twins fitted out with clothes. She did not want a strike but her support of my grandfather's actions was absolute. Her view was that he knew best and that it was his right to decide how best to cope with the strike. The 'Workers' Chronicle' of the Newcastle Trades Council of Action singled out women for special praise in its eleventh issue:

> One of the most encouraging features of the present crisis is the glorious spirit shown by our women folk.
> Everywhere they have thrown themselves wholeheartedly into the fight. At the mining centres we see them active encouraging the pickets to do their work thoroughly. Where feeding centres have already been set up, there we find them toiling merrily all day long. With the active support and help of the women we can go forward defiantly to the conquest of Capitalism. (Gateshead Public Library)

My grandmother's contribution was not obvious or outspoken; it was a calm determination to bear the burden of the budget and never to question my grandfather's reasons for sticking strictly to his union's decisions. As I shall show, although not obviously political, my grandmother's support, expressing itself in a creative willingness to muddle through, to make do with nothing and to scratch resources together as best she could - and she was helped in this by the children - was central to the family's ability to battle through the eight months which followed and during which time they had no income whatsoever. Her experience of class conflict, as it were, in the kitchen was just as sharp as my grandfather's in the pit.

What worried her in the early stages was the court case hanging over Jim. And she worried about Olive, too. Olive's husband, Tommy Willis, was a pitman and he was also on strike. By this time Olive had two daughters, Francy and Sally, and they were as poor as church mice. My grandmother often dispatched my mother to Crawcrook to help Olive with the babies and to take some vegetables or home-cured ham. Her aim, in short, was to protect her family as much as she could; she left the political worrying to others.

The strike ended messily and bitterly (see Mason, 1970). From C.R. Flynn's account it is clear that almost to the end victory was expected. When the reality of defeat began to dawn the reaction throughout the north-east was one of shock and anger. On 14 May the TUC's 'British Worker' tried to interpret the ending of the strike as a kind of victory:

> Fellow Trades Unionists, the General Strike is ended. It has not failed. It has made possible the resumption of negotiations in the Coal Industry and the continuance during negotiations of the financial assistance given by the Government. (No. 4,

1926, Gateshead Public Library)

The reaction of the Newcastle Trades Council was radically different and expressive of a much angrier mood of betrayal:

> Never in the history of working-class struggle – with the exception of the treachery of our leaders in 1914 – has there been such a calculated betrayal of working-class interests as has overtaken us this week. ('Workers' Chronicle', no. 14, 1926, Gateshead Public Library)

The collapse of the General Strike reinforced my grandfather's view, which he had expressed often enough in recent weeks, that 'strikes never did any good to anybody'. He was bitterly disappointed at the result. He was a man who never allowed his optimism to outstrip his common sense but he had felt, for a while, that the miners might win this time. That feeling evaporated early in May. What did not change was his determination that they should, none the less, fight it out. The collapse of the General Strike confirmed his suspicions that the miners' strike would be a long one and that the preparations they had made, as a family, had been justified. His weary recollections took him right back through time; 1921, 1912, 1887 were three critical dates in a legacy of hard struggles for minor gains and sometimes major losses.

THE LOCK-OUT

I have had several Throckley pitmen tell me that the end of the General Strike was a sell-out and that the miners were in effect betrayed. Their immediate feelings were those recalled from 1921, feelings of isolation and betrayal confirming their view that whatever the miners got they had to get by themselves. And what lies behind this is their great sense of pointless sacrifice. Bill Straker was to refer to the lock-out as 'probably the greatest industrial conflict known to history ... a heroic struggle' (NMA, Minutes, monthly circular, December 1926, NRO 759/68). And in the same circular he noted, 'Under the terms of settlement to which the miners of Northumberland have had to submit there will be thousands in a state of semi-starvation.'

But that was a view from the end of the lock-out; in the early weeks the outlook was less gloomy. In its 22 May edition the 'Newcastle Journal' noted (in one of a series of sketches titled In Pit Villages) that

> the third week of the stoppage finds the average miner apparently content to leave affairs in the hands of his leaders.... In the main yesterday's tour of half a dozen villages failed to find signs of despondency, although the talk of deadlock after deadlock is not inspiring.

The paper quotes one North Walbottle optimist as saying, 'The
last six months of a strike are always the worst.' Generally,
however, the paper described the miners as 'busy in their gardens or out on the football pitches'.

My mother recalls the early days of the lock-out by the
weather:

> I was 14 years old when the 1926 strike broke out.... As it
> happened it was May when the strike started so the days
> were light, bright and warm.... At first it was exciting to
> go along the Butcher bank to see the pit ponies. Some were
> blind, always been in darkness and then the sudden light.
> We often took crusts along to feed them.

Early in June the 'Newcastle Evening Chronicle' published a full
account of the situation as they saw it in the Heddon, Throckley
and Walbottle area. I reproduce the report in full since it does
evoke a rich picture and reveals, too, the implicit hostility of
the local press to the miners' case.

IN THE VILLAGES

How Throckley and Heddon View the Stoppages
Guardians in Reserve

The area comprising Walbottle, Throckley and Heddon-on-
the-Wall makes a convenient objective for an investigator
whose mission is to find out how life goes in the pit villages
during the sixth week of the stoppage.

As in the case of the eastern villages, Seaton Delaval and
Seghill, there seems to be little outward sign of industrial
conflict except that men are to be found at convenient cor-
ners and in the open spaces.

Nobody looks ill-fed, and, with one exception to be dealt
with hereafter, nobody looks despondent. In all three villages the inhabitants sat out in the sun yesterday and
thankfully accepted the cooling breeze.

At Heddon especially the children were plump and rosy,
and a stranger dumped down suddenly outside the Church
Schoolhouse would certainly have found little grounds for
suspecting there had been any cutting of rations.

In this happy little corner of Northumberland things have
gone very smoothly from the first day of the stoppage. Our
representative was told by three competent authorities that
during the days of the hold-up of traffic the Heddon district men kept well within the law.

There are, of course, a certain number of hotheads and a
few more or less 'humble disciples of Lenin', but these are
swamped by a majority of level-headed pitmen, good crafts-
men at their job.

No Rush For Relief

It may well be that if the dispute lasts a few weeks more
there will need to be a certain tightening of belts, but
although at the moment the aid of the Guardians has not
been invoked generally, the reserve is there when needed.

It was gathered that the feeling is that if appeals to the
Guardians can be avoided they will be by this independent
little community.

Throckley is slightly different in some respects. There
seems to be a little more militancy and a frequent and free
outspokenness when mining topics are on the wayside
agenda.

There is, however, the same desire to carry on without
undue 'grousing' although there may be a surplus of
denunciatory criticism.

Yet even here the miner, apart from his mates, is not
averse from acknowledging that the exporting collieries
must be vitally concerned with the incidence of foreign
competition.

In groups, of course, the men are likely to think in
groups. The chief hope seems to be that some means will
be found to maintain pre-stoppage conditions pending
reorganisation and the latter will make the pits flourish
exceedingly.

'Ower Mony Gaffers'

The chief 'grouse' voiced was that there were too many
officials underground or 'far ower mony gaffers' in the
vernacular.

Walbottle in general might be termed unconciliatory,
though other people might have other names for it. The
village is full of die-hards and die-oftens, if the expression
be permitted.

Herbert Smith is deified and Cook canonized twice nightly
and during the day. On the other hand, the general opinion held of the coalowners' officials need not be put down
in cold print.

Yet meeting and talking with the men singly one finds
in them a readiness to agree that nobody is likely to run
pits at a loss just for the fun of the thing, and that there
must be another side to the argument.

A Pessimist

There are plenty of humorists in Walbottle who take life
lightly even in these hard times, and there is one prize
pessimist. He was found leaning over the wall at the corner and seemed as if he had feared the worst since birth.
The following dialogue ensued:-

Investigator: 'Well, there's this to be said, the pitmen
can get in a store of sunlight during the dispute.'
'Aye and fowks winna' need coal and sae canna feel the
pinch syem as the miner.'
'But the pitmen are getting out into the air all day long.
Surely that's something?'
'Aye and gettin' a canny appetite and sae eatin mair.'
'Well, it can't last much longer and then the pits will
open and you'll be alright then.'
'Aye, but then we'll ahl hev te gan te work again!'
Verily some people are hard to please.

THE POLITICS OF PICKING COAL

The image of the miner at the gatepost is, however, a false one;
the lack of commitment to work which the reporter assumes is
quite misleading. For pitmen in the area the whole point of the
lock-out was that they were struggling to improve the conditions
in which they worked, and the image of people lazing in the
sun conceals the reality of all the mining families in the district
settling down into a protracted dispute and making preparations
to heat their homes, feed themselves and struggle through. My
mother's account of the lock-out brings this out vividly.

I remember my parents to be very upset and worried. The
coalhouse was stacked high with coal and wood, but the only
means of cooking and heating water was a coal fire, so stocks
gradually deteriorated. That was when Jack my twin brother
and I decided to look for coal. I heard of people going along
the Butcher bank picking cinders. I found the cinder plot so
Jack and I took the boogy and lots we got, (now they call it
coke). After that we discovered we could get coal dust from
the Maria pit yard. Again Jack and I went and gathered lots
of it; we made eggs with our hands; we soaked the dust with
soapy water. It was fun doing them, and very hot they were.
It solved the problem for a time, then my brothers and Dad
discovered the coal in the Dene. Then again Jack and I took
the boogy to the Dene side, and we pulled the bags of coal to
Mount Pleasant.

Many people did the same things, grubbing coal from outcrops
or even tunnelling in the Dene to get it. It was, of course,
illegal and the police tried to prevent it. This activity gave rise
to a key episode in the lock-out.
My uncles Bill and Jim were working in a dug-out in the Dene
and my grandfather was standing at the entrance, pulling out
the coal they had cut and keeping a watch for the police. A
policeman did steal up on them, however, and ordered them out
of the hole. My grandfather explained to the policeman what they
were doing, and that they needed the coal, but the policeman,

so my relatives report, was adamant and took a threatening attitude which angered the old man. The policeman told him that if
he did not get the lads out of the hole he would kick the props
away and let it collapse on them. My grandfather exploded and
told him, in such a threatening manner that my uncles were
startled, that if he went anywhere near the props he would kill
him. The policeman left, presumably to get help, and the men
left, too, their coal in a barrow and, if their account of it is
still to be believed, their suspicions confirmed that the police
were hell-bent on strike breaking.

This particular story, I might add, has now the status of a
family parable. I was told it in my childhood and have heard it
many times since; it conveys for my relatives all they feel about
the character of my grandfather, injustice, the struggles they
went through and the police. It invariably ends with the comment
that my grandfather always said he was too honest to be a bobby.

Picking coal also carried other risks, for the union stance on
the matter is clear. The Executive minutes for 26 May state:
'that we condemn the action of any of our members who have
produced, or may produce coal for sale, as such action is absolutely contrary to the general stoppage in which we are now
involved' (NMA, Minutes, NRO 759/68). My mother writes on this
issue: 'The lads often sold a bag of coal for a little pocket money.
If they got twopence for a packet of Woodbines they were happy.'
My grandfather did not disapprove of his sons fiddling a bit this
way and there is no evidence that the local lodges did either.

PERCEPTIONS OF THE POLICE

The dug-out episode and the arrest of Jim for picketing brought
into question the role of the police in the strike. Mistrust and
suspicion were conventional attitudes to the police, but the
accumulating evidence of police action in the area added a new
dimension, antagonism and a sense of frustration born of powerlessness. There was another episode, in fact, at the dug-out
involving my uncle Bill but not my grandfather. One police visit
the clothes of the men working in the hole, including Bill's,
were confiscated and never returned, a minor act of provocation.

Police strategy during the lock-out was not based on the
expectation of much trouble, although they knew from Lieutenant-
Colonel C.E. Maude of Northern Command that the troops were
ready to be mobilised if necessary (Chief Constable of Northumberland, File: Official Circulars re Emergency, NRO NC/1/20).
The general plan was to call up the police reserve to police the
agricultural districts of the county and to move the full-time men
to the mining districts. The chief constable wrote later of this
strategy that 'by these means it was possible to avoid any calls
on the military for any outside work' (Chief Constable of
Northumberland, Standing Joint Committee Minutes, 1926-8,
NRO CC/CM/S5).

Anticipating 'a movement by Communist Agitators to stir up trouble in towns and other places', the chief constable directed his superintendents to use shorthand writers at strike meetings (NC/1/20). And on 22 June he instructed the superintendents 'to place themselves personally in touch with the Managers of all pits in the division'. This was to help those men who wanted to get back to work to do so under police protection. In addition to intensified regular policing, minor harassment of the sort described and helping blacklegs, the police also sought to control meetings. On 3 November, for instance, the chief constable informed the Home Office that two meetings in the Throckley district (at Westerhope and Walbottle) had been banned. They were to be addressed by Will Gallagher. Two reasons were given for the ban under the emergency regulations:

(1) The poster announcing the meeting contained the words 'Come and join the C.P. and defend your rights against the Forces of His Majesty.'
(2) The probability of intimidation of miners returned from work by the presence of a crowd at this point was also a ground for prohibiting the meeting.
We have recently had trouble in this district.
(NC/1/20)

Police tactics in this dispute were orchestrated directly from the Home Office by Joynson-Hicks, the home secretary, and the use of the police in this way necessarily raised questions about whose side they were on, the impartiality of the law and the determination of the government to break the strike rather than find a negotiated settlement. The reputation of the police, such as it was, reached its lowest ebb and the miners' sense of being the victim of government policy was heightened.

The court case in which Jim was involved confirmed their suspicions. Oliver Akenside was fined and imprisoned for one month and refused the right of appeal. Uncle Jim told me that they were all sitting waiting to go into the court room, expecting only to be fined. 'Wey, lad,' he told me, 'when we heard what Akey got we got the shock o' wa lives. We aall thowt we would gan doon. Mind ye, we should a' done it allreet.' In fact, they were fined £5 and the union agreed to pay the fine. The chairman of the bench told them that he had taken 'a lenient view' ('Newcastle Evening Chronicle', 18 May), but things looked different from Throckley since they felt the police had misconstrued the evidence anyway and exaggerated the size of the crowd picketing. Oliver Akenside's imprisonment was thought to be totally unjustified.

FAMILY MOBILISATION AND FUN

While the trades council and the union pursued the struggle in
the political arena, ordinary families battled it out in the home in
sometimes desperate attempts to keep warm, provide food and
keep spirits up. In addition to collecting fuel some families
managed to get work on the local farms. My grandfather continued with his casual work on Lamb's farm, receiving payment
in kind. Even the children were mobilised to work. My mother
again explains:

> It was a hot summer. Our Jack and I got a job picking straw-
> berries at Mordue's farm at Heddon station. We had to be
> there at 6 a.m. to get the strawberries on the 9 a.m. train.
> All we got was 1½d. for picking a 3 lb basket. Our wage was
> around 6s. 6d. a week (two weeks picking) but that was 13s.
> a week for my mother and she was really grateful to get it.
> Then Jack and I went out potato picking, I think we did a
> month at that; two weeks at the colliery farm, and two weeks
> at Jim Hedley's at Heddon. We got 2s. a day for that and a
> pack of potatoes each day. That was a big help, the potatoes
> were big ones, so mother kept them to make chips. But our
> wages - I always remember her saying, 'I will straighten up
> my insurances with the money the bairns have earned.' She
> was worried in case her insurances ran out.

Her account goes on to describe vividly some of the preoccupations of the family and their different ways of coping with them
and, indeed, enjoying them.

> Our biggest worry was clothing and boot repairs. We often
> hunted for old tyres for Dad to cobble our boots as leather
> was impossible to buy, as we had no money. In fact, we
> liked the rubber on our boots. Concerts were given in the
> store halls, by the local talents. We enjoyed those nights out.
> We also enjoyed going to the Dene to gather wood. The big
> lads could saw the branches; we girls packed them in sacks,
> and brought them up to Mount Pleasant in the boogy. That
> poor boogy worked hard. It never once let us down. Some
> people pulled coal and wood on sledges and the lucky people
> had barrows. My Dad had a lovely big barrow; he made that
> work overtime.
>   Our Eva was at home, she had a little girl to bring up; her
> husband had run off and left them, so she applied for 'parish
> money.' She got 12s. for Olive and her; that went towards the
> housekeeping money. The stores sent the groceries every week
> (on a bill); that had to be paid for after wages came into the
> house [i.e. after the lock-out was over]. Our only means of
> getting pocket money was to earn it (go shopping to the
> Co-op, or clean someone's house for a shilling) someone that
> had a wage earner in the house.

I remember helping to clean Mr Reay's house, ready for
him to move to the Leazes, I got a few shillings for that.

My mother's account is not at all gloomy. She looks back now and
recalls how difficult things were, but is clear that at the time
life was not always depressing. In fact, the village was busy and
at times quite jovial. The men in the club organised walking
competitions during the summer to fill the time in and raise a bit
of money.

Albert Matthewson told me that in the top rows there were
regular evening running competitions among the young lads, and
they played a lot of football. 'Football teams', he told me, 'would
turn out as late as nine o'clock at night.' There were concert
parties in the store hall to raise money for the soup kitchen.
And there were occasional crazy football matches between men
and women or just between women themselves, all designed to
raise money for the soup kitchen. There were occasional chara-
banc trips to the seaside in the so-called Dollar Princess,
organised by the store or simply by a group of people themselves.

It is hardly surprising that my mother's account of these
days is tinged with a sense of pleasure; she was, after all, quite
young.

Fund raising was not confined to Throckley. Throughout the
country local Labour Parties organised help for the miners;
many local parties 'adopted' pit villages, directing their efforts
to a particular place. Throckley was adopted by the Hastings
Women's Guild of Service. They sent clothes and donations to
the Distress Committee (run by Dan Dawson) and the committee
sent them a miner's lamp to raffle. They sent shoes, too, to
help out the Throckley boot and shoe fund. In July, Danny
Dawson wrote to the organiser, Mrs Hickmott, thanking her for
clothing 'to help ward off the brutal attack upon their already
low living standards', and he gave her this thought:

When poor men's tables waste away to barrenness and drought,
There must be something in the way that's worth the finding
out. When surfeit one great table bends, And numbers move
along; While scarce a crust their board extends, There must
be something wrong. (Dawson Papers, NRO 527/B/1)

The Browns got nothing from the Distress Committee; in a
sense they did not need anything. As a family they worked
together; they had their gardens and pigs and a little money
coming in from Eva. By mid-summer Jack was adding twelve
shillings a week to the family budget.

Jack left school during the lock-out and managed to get a job
as an apprentice joiner under Mr Henderson, the local undertaker. It was not just work on coffins; the business included
work on local farms and houses. He did help with funerals. 'Our
Jack', says my mother, 'made a lovely undertaker; he was cut

out for it.' My grandmother had always worried that Jack might
have to start at the pit, and she felt quietly pleased that the
strike and lock-out had prevented this. They were all very
proud of Jack's achievement in 'getting a trade'. My mother says,
however, that she 'always felt sorry for Mr Henderson. His face
was bright red and his eyes were always tearful'. My grandfather assured her, though, 'They're not bloody tears: that's
the whisky come out of him.'

THE STERN FACE OF CHARITY

Beyond what families could do for themselves there were other
institutions in the village mobilised to battle through. These
included the Distress Committee which organised the soup kitchen,
the store which extended credit, the club which funded competitions, and even the district council. I shall discuss these briefly
in a moment. It is important at this stage to realise that if
these resources failed the people would have to fall back on the
Poor Law and the Castle Ward Board of Guardians. The Castle
Ward Board of Guardians dispensed out-relief to claimants in a
regular way; miners, on the other hand, were given money on
loan. By August, the payments in out-relief were four times
the original estimate and the secretary to the guardians,
Mr C.S. Shortt, recommended that payments on loan to miners
should cease. 'If the Board desires to comply with the law, all
relief on loan to miners should cease' ('Blaydon Courier',
7 August 1926). He went on in his report to give an account of
the board's recent finances: 'Out-relief since commencement of
the half year, £21,500 i.e. weekly average of £1,104. Original
estimate for same period £5,400 and £300.' Mr Shortt urged the
board to consider that, in contrast to the days when trade unions
maintained their members on strike from their own funds,

> to-day, Guardians throughout the whole country, when trades
> union funds are exhausted, have adopted the practice of
> granting to men engaged in an industrial dispute sufficient
> out-relief on loan to finance the trade union indirectly.
>   In other words, Guardians elected by and representing the
> general body of ratepayers use the rates levied on all classes
> including the employers involved to support sections of the
> community in industrial disputes which devastate and must, if
> allowed to continue, ultimately destroy the whole country.
> ('Blaydon Courier', 7 August 1926)

He went on to point out that the miners could be dealt with
under the Vagrancy Act. Miners read this sort of thing in the
middle of the dispute, and it is hardly surprising that they
detested the Poor Law and avoided it if they could. In this context the Throckley resolution to the council of the union in May
'that a further grant of £10,000 be set aside from the funds of

our Association towards the relief of our unemployed members in
the county' takes on an entirely new significance. Miners were
determined to be as self-reliant as possible; the irony was, of
course, that their union funds could not stand the strain of it.
An NMA survey of branches in July showed that up to that point
no guardian payments for the relief of miners' children had been
paid out in Throckley (NMA, File on Replies to Questionnaires,
1924-33, Burt Hall).

The lurking threat of a demeaning application to the guardians
for 'parish money' accounts, too, for the way in which the store
and the club, the Labour Party and even the District Council
acted during the lock-out. For these institutions, too, were
where possible mobilised for the conflict.

THE UNION JACK CLUB

The club was an important meeting point for the men although
its drinks sales dropped. Here they could talk among themselves
and assess the situation, and there were, in addition, concert
parties and competitions. The club played a small role in directly
supporting the miners. On 1 May they protested to the prime
minister about his handling of the miners' case (Throckley Union
Jack Club, Minutes of Committee). On 29 May they gave all
paid-up members a £1 voucher to buy goods from the store.
They donated money to the Sports Committee to finance walking
competitions, and on 12 September paid £1 to the Distress Committee. They donated money to the boot fund, and in October
resolved that 'no member who is considered a Blackleg will be
allowed to play in the Tournament games'.

THE CO-OPERATIVE STORE

As in 1921, the store extended credit to its members. 'The
co-op', Albert Matthewson told me, 'helped a lot.' People withdrew their savings or ran up debt. They were, he said 'too
proud to draw assistance'. Average credit per member went up
from £2.1 in 1925 to £3.2 in 1926 and £3.5 in 1927. Average
annual purchases dropped from £45 in 1925 to £41. The policy
of the store was not to allow people to get too much into debt,
and to allow them to pay back their debts by not claiming the
dividend they were entitled to. Although the figures for the
strike and lock-out period do not indicate massive changes in
purchasing habits, it has to be remembered that throughout the
1920s both sales and profits in the store were falling. Average
purchases per member in 1920, for example, amounted to £93.
Additionally, the overall figures conceal a massive shift in the
pattern of purchasing. During 1926 the sales figures for the
butcher's and draper's departments were almost halved. Meat
purchases dropped from £11,837 in 1925 to £6,488 in 1926; the

respective figures for drapery are £6,755 and £3,237. Families
cut back severely in their meat consumption and their renewal
of clothes.

Store credit is the topic for another Brown family parable
which I have heard many times. Long after the lock-out was over
my grandmother was called into the store manager's office to be
told that she had successfully paid off her debts. As the story
is reconstructed all the details are richly filled in, the dialogue
is reported, and the tale is told to emphasise the moral rectitude
of my grandmother. She made certain that she paid her debts
and was intensely proud of having done so. It is the old theme
of respectability again. And it is always told against the 'knowledge' that many more never repaid their debts.

POLITICAL AND INDUSTRIAL TACTICS

In a sense, what I have been discussing up to this point is how
people in Throckley, my own family in particular, coped with
the conflict situation they were in. The conflict itself, however,
was directed and given focus by the union and by the politicians.
The Labour Party was not very active during either the General
Strike or the lock-out. Indeed, the minutes of the Wansbeck
Divisional Labour Party for the whole of 1926 do not mention the
miners' struggle. Some of the officials were, of course, active in
other ways, in the union, on the district committees. The Newburn and District Party was a little more active and arranged in
July for Arthur Cook, the secretary of the MFGB, to address
their annual meeting. This visit is an interesting one for my
account; it reveals something of the political mood of the miners
and my grandfather's attitude to the union leadership. During
the lock-out Cook undertook a prodigious programme of speaking. On Monday, 3 August, he addressed a meeting at Walbottle,
just along the road from Throckley. The 'Newcastle Journal'
described the speech, and the meeting at which it was given. I
reproduce the report in full:

MR COOK WANTS A SETTLEMENT.

More Extravagant Speeches.
Venomous Attack on Premier.
Police Threatened.

'Cook's tour' - to adopt the phrase of Mr A.J. Cook himself
was continued in Northumberland yesterday when the secretary of the M.F. visited Prudoe and Walbottle.

At each centre he found a crowd of men, women and
children approx 10,000 in strength and an abundance of
red favours in evidence; whilst the red flag was played
and sung.

Introducing himself as the 'villain of the piece' he

encouraged his followers to wait for the victory he pre-
dicted, and added a new lash to his whip for the benefit
of the police against whom he uttered a threat. He made
a venomous attack on Mr Baldwin.

No Surrender! Speech in Drenching Rain

A throng of supporters besieged Mr Cook's car when he
arrived at Walbottle, and he was carried shoulder-high to
the platform. Rain commenced to fall as he addressed them,
but he continued until he and his audience were drenched.
    'I want a settlement', he admitted. 'Who doesn't?'
    'But there is going to be no surrender. You stick it:
we win. If you can hold out and we can feed the children, victory is as sure as the rising sun.'
'The influence of the commercial world will compel the
owners very soon to open the doors to negotiations which
may lead to a satisfactory settlement,' he added.

District Settlements.

'Does it look as if we are beaten when our delegates are
touring all parts of the world to collect funds to help us?
The only way we can be defeated is by the district settlements, and it will require all the power of local committees
to prevent them being accepted.'
    Mr Cook expressed himself 'sick and tired' of different
organisations pulling different ways, and said the great
need was for one large Federation. He was prepared to
recommend a national ballot of the miners if Mr Baldwin
would ballot the nation.

The rhetoric was powerful, and although none of my relatives
was at the meeting they certainly heard about it. Uncle Jim thinks
that Cook was a fine man but my grandfather did not trust him.
He felt that Cook was unpredictable and, despite his rhetoric,
likely to sell the miners out on almost any terms. This view is
partly a reflection of Cook's known opinion that as the lock-out
continued that the miners might have to settle in the short run
for unfavourable terms aiming, in the longer term when they had
greater strength, to win their battle decisively. But it was more
firmly based, I think, on my grandfather's assessment of Cook
as a man and as a leader.
He 'knew' that Cook was a communist and something of a political firebrand. He knew, too, that Cook was a powerful speaker
and that he had been imprisoned for his views in 1918 and 1921.
Cook, in short, represented class politics, something my grandfather was greatly suspicious of. Paradoxically, it was the
powerful upsurge of class feeling in my grandfather as the lock-
out continued which turned him against any settlement in which
the miners might lose, and which therefore reduced his

confidence in Arthur Cook.

But there were other factors; Cook was an MFGB man, a
Welshman and highly political. My grandfather had been schooled
in a union which favoured the sober politics of parliamentary
reform and district agreements in industrial relations, and in
which union leaders had been pillars of respectability in their
own communities and in the union as a whole. Cook was clearly
a leader of a new kind. And his respectability was brought into
question in a direct way for my grandfather.

A.J. Cook stayed overnight in Throckley with Danny Dawson.
Apparently, his shoes were so badly worn and wet that they had
to find him a decent pair and discard his old ones. Apocryphal or
not, this story found its way back to my grandfather and it confirmed his view, at least in my uncle Jim's opinion, that so far as
A.J. Cook was concerned 'there was something not reet': leaders,
it seems, had to be above reproach in their personal bearing and
affairs if they were to command his respect. Cook had clearly
failed the severe test my grandfather applied to everyone in
positions of authority, the test of respectability.

He might not have respected Cook as a man but he did share
his views of the lock-out. My grandfather never doubted the
wisdom of sticking it out and in this he took his cue not so much
from the politicians as from his union. One final point about politics: the lock-out strengthened the commitment of many miners to
the Labour Party. The Labour Party in the district had studiously
avoided any working alliance with the Liberals in the course of
the dispute and looked forward to a Labour government as a way
of solving the coal crisis. The passion and character of some of
these beliefs comes out rather well in a letter which Danny Dawson sent to Mrs Hickmott of the Hastings Women's Guild of Service
on 2 November thanking her for the gift of clothing which had
been sent. The clothes were welcome

> as everybody is getting down very badly on clothing, and the
> weather is bitter cold up here. In spite of this, the spirit and
> determination of our people is splendid in resisting the despic-
> able tactics of the coal owners in their endeavours to force
> them back by hunger and starvation, to conditions unknown in
> this County for 80 years. And they are faithfully backed up by
> honest ? Baldwin and his followers. (Dawson Papers, NRO)

He went on to express a hope that 'we will soon have a Labour
Government administering the affairs of this country and breaking the power of the Dukes of Northumberland, the Earl Percys
and Baldwins etc etc'.

BLACKLEGS

The union struggle was carried out at three levels, nationally,
at county level and by local branches and trades councils. These

levels were not always well co-ordinated or even in agreement.
In October, for example, the council of the NMA passed a resolution favouring compromise in negotiations with the coal owners,
but the branches overwhelmingly rejected the idea. From August
onwards, however, the union was actively trying to persuade
men not to go back to work. The local press took great delight
in publishing news of some pits returning to work. On 18 August
men started again at Walbottle colliery, Wylam and East Walbottle.
On 1 September North Walbottle started again. On 28 September
the 'Newcastle Journal' reported that over 1,000 miners were
back at work in Durham. On 2 October the 'Journal' ran the
headline Miners Flocking Back. It was this drift which prompted
the union to call for the compromise in negotiations. But when the
call was rejected the following circular to branches was issued
urging the men to stand firm:

> Wages will be reduced to only one-fifth, or 20 per cent,
> higher than before the war, while the cost of living stands
> at nearly three fourths or 74 per cent higher. Under these
> conditions, even the common necessities of life will not be
> possible for you. Once there is a settlement the demand for
> coal will be great and the price will be high. Profits will be
> large, while wages will be lower than the merest poverty
> line.
> We appeal to you to stand firm until there is an honourable
> settlement. The struggle has been hard and long, but not so
> hard and long as the slave conditions and wages offered by
> the coal owners, if once conceded.
> Stand loyally by your association. Without it what will the
> future be? Mines Regulations Acts, Compensation Acts, all
> will be altered at the will of the coal owners and other great
> employers. You have everything to gain and nothing to lose
> by your loyalty. ('Newcastle Journal', 25 October)

The problem of loyalty at the local level was essentially one
of blacklegging (see Moore, 1974). There was a big demonstra-
tion by many Throckley men against the return to work of men
at North Walbottle pit and the Coronation pit at Blucher. Stones
were thrown at men leaving work. The size of the blackleg
problem was not great – as late as 1 November the NMA estimated
it at 5,000 men out of a workforce of 56,000 – but it was a very
emotive issue. There were near riots all over the county over
blacklegs; there was trouble at Ryton just over the river and at
Silkworth, Durham, there were episodes in which police made
baton charges into crowds demonstrating against blacklegs.
Among the Throckley group of collieries there was trouble at
Blucher pit where 200 men – not members of the NMA and
annoyed that they had received none of the so-called 'Russian
money' or 'Russian gold' which had been received elsewhere –
opened up negotiations to start work. That was in early June
and many Blucher men (thirty-five in all) did go back. In

Throckley a group of blacklegs tried to form their own union, the so-called Blacklegs' Union, but the attempt failed.

Blacklegging gave rise to an issue which affected my grandfather for the rest of his life. His 'marra', Mr Guthrie, under pressure from his wife, so it is said, went back to work. He spoke to my grandfather about it, saying that he could see no way out of his problems. Unable to persuade him to change his mind, my grandfather could only explain that he could have no more to do with him, and to the end of his life he never spoke to George Guthrie again. Long after the troubles were over the issue of blacklegging still created ill-feeling in Throckley. Albert Matthewson told me that 'to be honest, bitterness still exists. It's never been forgotten'.

DEFEAT

The defeat of the miners was a slow one and messy in that a growing number of pits resumed work throughout October and early November. The collapse of the will - even, perhaps, the ability - to go on occurred mostly in the last week of October and the first fortnight of November. On 16 November the council of the NMA discussed proposals to carry on with the fight, but concluded:

> The Conference had before it the number of men who had returned to work in the various districts as near as could be estimated by the reps. The position in almost every district was most unfavourable to fighting on, so that it was seen that fighting on was out of the question. (NMA, Minutes, NRO 759/68)

On the recommendation of the MFGB, negotiations with the owners were opened up on a district basis. These negotiations produced an agreement involving a reduction of one shilling per shift, the establishment of three shift systems in some areas, and a two and a half hour extension to the working week. The Executive of the NMA referred to these as 'slave conditions'.

To every miner the conditions represented a defeat. The hardship of the lock-out had led to nothing but disaster. The MFGB was practically bankrupt and the unity of the union was threatened by the establishment of company unionism in some coalfields, Nottinghamshire in particular. The defeat of the miners left deep political wounds in the labour movement. Margaret Cole has argued that 'what really perished in 1926 was the romantic idea, dating from before the First World War, of the power of syndicalism, direct action and the rest of it' (Cole, 1977, p. 14).

Arthur Cook, who had warned that if the miners were driven back to work they would start a guerrilla war underground, issued a pamphlet at the end of the lock-out referring to the way in which some leaders of the labour movement - Philip Snowden,

J.H. Thomas and Arthur Pugh - had supported the ending of the dispute. In this pamphlet he revealed, quite deliberately, just how deep the wounds were in the labour movement.

> Judas, at least, had the decency to hang himself in Alcemada. He did not write articles recommending peace and co-operation with Herod and the Romans; that work he left to the scribes and pharisees. There is no miner now, no miner's wife, no miner's child above the age of ten or younger, who does not know that in this great fight the people who helped to starve them were not only the coal owners, but the policemen, the magistrates, the Boards of Guardians, the Cabinet ministers, and the whole array of the state, both in the national organs and in its local organs, and that in addition even against all this force they would have won had it not been for the open treachery of some Labour Leaders who should have been their friends. ('Newcastle Journal', 27 November 1926)

The sense of betrayal was acute; it was compounded by uncertainty, for although the terms for going back to work had been agreed it was not clear whether work would be available or who would get it. Christmas was approaching and nobody had any money. The Brown children got no presents that Christmas although they did feed well. In the dark days at the end of the year my mother remembers them having to be very sparing with coppers for the gaslight and being very short of coals. My mother's assessment of it all now is that 'our young lives were spoiled by strikes but the experience made us thrifty ... we knew how to manage properly after being short so long, a lesson we will never forget'.

Most accounts of the lock-out are rich in their descriptions of hardship. Some, too, emphasise that hardship was borne and experienced in different ways and to different degrees within the mining community itself (see Noel, 1976), and I end this chapter with a comment by Dr Messer, the Newburn MOH. Writing in his annual report for 1926 he observed:

> Undoubtedly the industrial depression and the crisis in the mining industry have influenced the infantile mortality. The chief causes of death were congenital malformation, debility and premature birth; diseases of the respiratory system and diarrhoea and enteritis. It is difficult to estimate the effect of the recent crisis upon the population generally. There is strong evidence that school children in the majority of cases did not suffer but in many instances actually improved.... Up to the present there is little evidence that the males suffered at all, but it would appear that the women folk and the infants born towards the end of the year did. (MOH, Newburn UDC, Annual Report for 1926)

In the following chapter the consequences of the lock-out on social relationships in Throckley and the pits are briefly discussed, the discussion being an introduction to the more protracted struggle the miners faced through the crisis of the 1930s.

11 THE DEPRESSION YEARS AND RETIREMENT, 1926-36

In the last chapter I attempted to describe something of the experience of industrial struggle. The form and course of that struggle were clearly related to the class position of miners and the conflict situation into which the economic policies of the government had manoeuvred them. Lacking the resources and cohesion and, therefore, the power to wage a more protracted fight, the miners were defeated. The tactical battles of the conflict - the press campaigns, the negotiating positions and so on - were a product of decisions contingent on the unpredictable developments of the lock-out and the perceptions of the miners' leaders. The underlying reality of the market situation of the men, the economic base of the industry as a whole, was, however, what in the end defeated the union.

In this chapter, bringing my account of my grandfather's life up to the point when he left the pits, I want to examine how the class position of miners changed in relation to the shifting fortunes of the industry in the years immediately following 1926. The focus of class conflict in Britain shifted during these years away from the coal industry, although not entirely. Unemployment, poverty and depression, particularly in the distressed areas, gave class conflict a new meaning and new political forms. Paradoxically, the period of the 1930s was also one of rising living standards, technological change and structural change in the economy (see Stevenson and Cook, 1977). But in the coal-mining areas the situation was different; phrases like 'the hungry thirties' remain pertinent. One of the essential hallmarks of a class society is that rewards and opportunities are distributed according to power rather than need. The experience of mining communities during the depression exemplifies this only too well.

THE RETURN TO WORK

The end of the lock-out was not for many men the end of being idle. The Throckley coal company acted cannily throughout December, taking back only those men they wanted. Some men were never taken back. My grandfather and Jim waited till the new year before they went back, the old man defiantly insisting that he would wait till he was called. He would not go back and beg for a job. His brother, Bob, was never asked back and had to leave his colliery house. He moved up Newburn Road to a council house and eventually to another job at Stargate colliery over the river at Ryton. Uncle Bill told me that they just waited around 'kicking their heels' to see what would happen.

There is some evidence that the coal company favoured employing blacklegs and non-union men. The replies to the Northumberland Miners' Association questionnaire to branches show that at Throckley Isabella pit twenty-two men had not been reinstated and that their places had been given to 'strangers'. In the Maria pit, the pit with the strongest lodge, fifty men were not reinstated. The report goes on: 'Twenty of the Throckley men reinstated were blacklegs and both branches reported that preference had been given to blacklegs' (File on Replies to Questionnaires, 1924-33, Burt Hall).

Bill was the first of the Browns to go back. He told me it was very difficult. The pit was in poor condition; the atmosphere was very bad among the men with those known to have black-legged being treated as pariahs:

It was rough going back after 1926 because the pit was rough as well. You were finding falls here and there and you had to dig your way in. I was one of the first of six putters back at the Maria and you had to go in and work with the blacklegs then.

As Bill said this his voice quivered bitterly and he pointed out that he was told by his father that on no condition should he co-operate with blacklegs. My grandfather and Jim were called to the pit after the turn of the year, after Bill had suggested to the manager that he should contact them. For a while, however, they both contemplated moving out of the district to Lynemouth colliery where they had heard men were being taken on. This was the one and only time that my grandfather considered moving out of Throckley.

A national trend after the General Strike of a fall in the number of members of trade unions was experienced in Throckley, too. Interestingly, there were differences between the two Throckley pits, with the Maria showing a much higher level of union membership among those remaining employed than the Isabella. But the Federation of Throckley Collieries, i.e. the joint committee of the branches of the three pits (Isabella, Maria and Blucher), was in disarray. The Maria would have nothing to do with Blucher because so many men at Blucher had blacklegged during the lock-out.

In both Throckley pits, there were immediate problems of unemployment which became more severe during 1928 and 1929. I estimate that for the urban district as a whole the proportion of the insured population unemployed in 1927 was around 35 per cent. Until the time my grandfather left the pits, unemployment in the district was rarely below 20 per cent although he himself was lucky to remain in work. The figures for the period are shown in Table 11.1

Table 11.1 Employment, unemployment and unionisation,
            Throckley Isabella and Maria collieries, Newburn
            UDC and Northumberland Coal Industry, 1926-36

	Throckley Isabella colliery			Maria colliery			Newburn UDC	Northumberland coal industry
							%	%
	(a)	(b)	(c)	(a)	(b)	(c)	unemployed	unemployed
1926	197	57.0	–	666	85.7	–	28.2*	–
1927	307	–	–	562	89.6	–	35.2*	–
1928	362	50.8	90	566	78.4	74	23.1*	–
1929	378	42.5	50	557	74.4	60	18.3*	–
1930	399	63.4	32	530	85.0	44	21.8*	15.7
1931	409	63.4	16	540	87.5	44	28.1*	21.3
1932	478	62.7	16	525	82.7	50	29.4*	23.4
1933	458	77.0	52	507	76.6	30	27.3*	19.0
1934	483	74.1	19	521	68.9	10	25.8*	16.9
1935	511	76.1	32	528	55.5	11	25.0*	14.8
1936	445	75.3	41	535	71.5	–	20.0*	–

(a) = number employed.
(b) = % in union.
(c) = number unemployed.
 * = estimated figure worked from projections backwards
      from December unemployment totals.

Source: Department of Employment unpublished statistics:
         Northumberland Coal Owners' Association, Statistical
         Information; NMA, half-yearly balance sheets; Mines
         Inspectorate, 'List of Mines'.

In the first month of the return to work the Executive Committee of the NMA agreed to finance a special pamphlet for the
Throckley pits to persuade non-unionists to join the union. In
February the Executive threatened legal action against some of
the hewers who had blacklegged for refusing to pay their share
of the checkweighman's wages. And in April there was trouble
in the Throckley collieries over the management's refusal to
deduct welfare fund levies from the men's wages. This was clearly
a management device to maintain rifts among the men. Some of
the non-union men had objected to these levies being taken without their written consent, and the reason was directly connected
with the fact that Danny Dawson, although no longer an employee
of the pit, was secretary of the Miners' Welfare Fund Committee
(NMA, Correspondence re Welfare Fund, Burt Hall). The company's argument was that this committee was no longer representative of the men.

Similar difficulties occurred later with respect to the Aged
Miners' Coal Fund which the Throckley federation operated.

This fund, financed by weekly levies, paid for coal for aged miners. In 1930 the company refused to deduct the levy without the written consent of the men. This infuriated the coal fund committee and their response to the company indicates just how far the reputation of the company had slipped. Pointing out that the sums involved were 'so trifling no man could frame a reasonable excuse to oppose it', Bob Butterwell, the secretary, went on: 'Speaking personally I feel the men want to get back to the old time when the Throckley Coal Co. was used as a reference for good through the county' (Aged Miners' Coal Fund, Minute Book, Tyne-Wear Archive, 1059/1).

Bill Straker summed up the situation of the return to work in the early summer of 1927. In response to suggestions from the Mining Association that the return to district wage settlements had been smooth, he noted in his April circular, 'The fact is that our people, after the terrible struggle of seven months against lower wages and longer working hours, were crushed and broken by starvation' (NMA, Minutes, NRO 759/68). And in June he noted:

> The present depressed state of the industry, with so many miners seeking work and not able to find it, has given a temporary power to colliery officialism, which I am sorry to find in too many cases being used in such a tyrannical way as to create the impression that they are glorying in the power to tyrannise over their fellows.

In Throckley such officialism was experienced in the way men were taken on or not taken on, over the levies question and over questions about paying minimum wages. The company was quite clearly prepared to respond to wage disputes (and these, as I have explained in an earlier chapter, arise all the time underground because of varying working conditions) by dismissal. In September 1928, for example, George Carr, the Maria branch president, was sacked for claiming county average wages for working in an abnormal place.

Such actions did not always meet with a passive response. In August 1927 the Executive Committee of the union had to counsel the putters at the Maria pit not to intimidate the management into paying minimum wages through threats to stop the pit. Anger had to be tempered with realism for, as Straker pointed out to the council of the NMA in 1928 when they discussed the possibility of industrial action over the Dodd's pay award, 'the men were in a much worse position to successfully strike than they were in 1926'. The feeling that nothing could be done was, perhaps, the most disheartening result of defeat, and what it gave rise to was not so much militancy as defensiveness, as I shall show a little later when I discuss the position in the early 1930s.

THE MINING INDUSTRY AFTER 1926

For the ten years after the General Strike the coal industry in
Britain was in decline. Stiff competition from Polish and German
coalfields, with their thicker seams and large coal cartels, are
clearly part of the reason. But the growing industrial recession
itself, affecting the internal demand for coal, was another. The
market position of the industry was adversely affected, too, by
insufficient mechanisation and high costs arising from the industry being dominated by too many small and, among the older
undertakings, inefficient pits. In 1929, 17.8 million tons of coal
produced in Northumberland and Durham were sold on the home
market. By 1933 the figure was 12.3 million (McCord, 1979,
p. 216). The collapse of the export market was even more
dramatic. In 1929, 21.0 million tons of coal were shipped from
the Tyne to foreign destinations. By 1932 this had dropped to
12.6 million tons.

Unemployment among miners was the direct result of these
complex causes and the underlying reason for their industrial
weakness. And unemployment was made worse by mechanisation.
By 1929 almost 55 per cent of the coal cut in Northumberland
was being cut by machine. In some of the newer pits sunk in the
northern part of the district the figure reached 80 per cent
(McCord, 1979, p. 216). By 1930, however, neither Throckley
Isabella nor the Maria collieries had conveyors, although some
faces had been turned over to longwall methods of coal getting.
In Throckley, then, the decline in employment was due to the
collapse of markets and to the inability of a small, ageing company to modernise.

Government policy for the mining industry in the years after
1926 was to encourage amalgamations in the belief that by so
doing costs could be reduced, output increased and profitability
restored. Kirby sums up the political dilemmas of this approach:

> In reviewing the Baldwin Government's policy on the coal
> mining industry between 1926 and 1929 the outstanding fea-
> ture is the conflict between the perceived need for ration-
> alisation and the political objections to any policy innovation
> which could be interpreted as a move towards public owner-
> ship. (Kirby, 1977, p. 121).

Legislation by the second Labour government - the Coal Mines
Act of 1930 - and subsequent attempts to help the industry to
centralise its marketing structures were rendered ineffective
by depression and the structure of ownership in the industry.
For such legislation did little to counteract the long-term structural problems of secular decline. The miners, in any case,
maintained their argument that only nationalisation could do anything for the industry. The opinion of the Northumberland
Miners' Association was that, despite the difficulties of the industry, output per man-shift had been increasing while wages had

been decreasing (NMA, Minutes, 1935, NRO 759/68). Between
1923 and 1934 wages per man-shift fell from 9s.10.39d. to
8s.8.56d., but output per man shift went up from 17.52 tons
to 25.17 tons. The number of man shifts worked fell from
nearly 16 millions to just over 10 millions. The variation among
different coal companies was, of course, quite large. The
Throckley group of collieries, being older and less efficient
than some of the pits, suffered higher levels of unemployment
(see Table 11.1).

But this period of contraction and cosmetic legislation is
important in a different way. The attempts of the national
governments of the period to intervene in the industry are part
of a much larger shift in the character of British politics and
society - the increasing involvement of the state in the management of the economy. The steps may have been hesitant and ad
hoc, but set alongside other changes, such as the growing
acceptance of Keynesian theories of demand management by
conservative politicians, steps towards regional planning under
the Special Areas Act and so on, these moves in the coal industry
were contributing to the building up of new relationships between
the state and the economy.

DEFENSIVE TACTICS IN THE PITS

The depression in the coal trade which persisted till the late
1930s shaped directly the industrial tactics of the miners by
neutralising their most potent weapon, their ability to strike.
Lodge records for the Throckley pit exist for the period from
1932 onwards and they reveal many of the weekly wrangles with
the company over a large number of questions, far too numerous
to list, but including such things as cavelling, dismissals, the
quality of house coals, prices for different classes of work, coals
for unemployed members and the problems connected with non-
unionism.

But what dominates the lodge discussions is unemployment.
The response of the Throckley lodges was to strengthen the
membership and to attempt to regulate how the available work
should be shared. A strong sense of locality and community
emerges from these efforts, the lodges seeking to protect the
employment of Throckley men against those defined as 'strangers'.
This question was first raised in March 1933 with the Isabella
lodge discussing 'strangers starting at the pits and our own men
still idle' (Throckley Isabella Miners' Lodge, Minutes, 1933). This
was a problem at the Maria, too. Charlton Thompson remembers
a lot of grumbles about the appointment of a new manager from
Haltwhistle, for not only was he a stranger but he brought
several men with him, some of whom took 'staff jobs'.

In 1936 the issue of strangers took a new turn. The minutes
read:

Secretary reported that he had interviewed the manager with
respect to strangers starting at the colliery and our own men
not working. When he said this man had started because he
had two lads at the colliery *and he was short of boys* and
was prepared to consider others unemployed in similar position raised by secretary. (Throckley Isabella Miners' Lodge,
Minutes, 23 July 1936; my emphasis)

This is a revealing comment. Not only does it indicate the
stronger position of management in avoiding normal cavelling
procedures for taking on men, it illustrates a subtle but substantial change in Throckley itself. Some of the older miners
still cling to a collective image of themselves as Throckley mineworkers with special claims to employment at their own pit. But
some of the younger men are unwilling to follow their fathers
underground. The cycle of the reproduction of a labour force
was breaking down rapidly in the 1930s, transforming the
character of Throckley itself from a mining village to an industrial community.

Charlton Thompson agrees with this interpretation. He told
me that his father was determined to see him down the pit,
telling him, 'I'll find you a job when you're ready', but Charlton
wanted something else. 'Lots of lads', he told me, 'didn't want
the dirty boredom of it. I think there was a turn away from it.
Quite a lot were getting work on the buses and different industries.' The irony is, then, that as the economy pulled out of
recession after 1934/5 and the demand for coal increased, the
social base of mining, the mining community, was disappearing.

This recruitment problem had its reflection at 177 Mount Pleasant. Jack, as I explained, had got work as a joiner. But in
1931, in an act of courageous defiance of his father, Bill, who
was worried about his health because he was recovering from
pneumonia, packed in his work at the pit, announcing to his
mother as he took off his clothes after his last shift, 'There, you
can hoy these in the fire; I divvent need them any more.' The
basic failure of the pit to capture Bill is illustrated not so much
by the fact that he finally left it but by the sensitivity which he
showed towards its brutalising and dangerous features. Bill
never felt easy about the pit; he could not take it or anything
about it for granted. He detested his pit clothes, finding them
coarse and itchy; he detested the management, finding them
more concerned with coal and machines than with the men they
employed; he detested the cramped, artificial world of underground. He told me once that he couldn't bear the 'awful feeling'
of the pit cage, after its ascent, dropping back for a split
second to settle on its catches. It wasn't so much the feeling in
the stomach that terrified him, as the thought that if those
catches were ever to break the whole thing would go crashing
back down the shaft. The likelihood of this happening was
remote but he says he felt it every time. Bill left the pit to be
unemployed, filling in his time in the gardens, taking on two of

his own at the bottom of Mount Pleasant.

For those remaining at the pit unemployment and short-time working remained possibilities. The policy of the Isabella lodge was clarified in 1933 in a six-point plan prompted by the dismissal of twenty workmen in July.

A. The sharing work with employed and unemployed effective at and including the previous dismissals.
B. That all non-union men should be paid off first.
C. That should any men be required cavils be put in for all men affected.
D. That they should have first claim to any temporary work such as filling small coals etc.
E. That all present vacancies should be cavilled for among the last batch of men who finished.
F. Abolition of any overtime (if any) that may be worked at the colliery when we have men unemployed.
(Throckley Isabella Miners' Lodge, Minutes, 31 July 1933)

The management did not agree to any of this and insisted that where only a few men were required they would choose whom they wanted, agreeing to cavel for men only when 'a bulk of men were required'.

IMAGES OF POVERTY

In its electoral appeals in the 1950s the Labour Party could use the slogan 'Just ask your Dad'. The aim was to evoke the image of the 1930s as the hungry thirties, the 'Devil's decade' or - another popular one - 'the wasted years'. The picture of unemployment, poverty and waste central to these images has its counterpart in much of the social commentary of the period itself. J.B. Priestley's 'English Journey' (1934) paints a bleak picture of the north-east. 'A nightmare place' is how he described it (p. 289), with Tyneside as 'a warren of people living in wretched conditions, in a parody of either rural or urban life, many of them without work, wages or hope ...' (p. 318). John Newsom (1936, p. 109), reporting on unemployment in County Durham, wrote of 'apathy and despair'.

The civil lord of the Admiralty, reviewing the problems of the special areas, said of the north-east, 'the area as a whole is losing hope' (Ministry of Labour, 1934, p. 74). The Pilgrim Trust report, 'Men Without Work' (1938), was a little more optimistic. The investigators found less 'desperate poverty' in the northern mining town of Crook than they had found elsewhere in the country, and 'a great determination not to give way to unemployment and not to subsist on self pity' (ibid., p. 74).

These descriptions reflect a shock reaction on the part of people from the south of the country on encountering directly

the problems of the depressed regions. There was a reality to
what they describe but their perception was selective for, to
convey a fuller picture of the experience of depression, it is
necessary to show how people fought back. Unemployment and
overcrowding, the Means Test and poverty were not just
accepted. The fight against them, through the Labour Party
and the trade unions, was what built up, during the Second
World War and afterwards, a positive commitment to a very different kind of social order.

The degree of poverty in Throckley is impossible to measure
now; the data do not exist. Poverty in the 1930s was, however,
closely associated with unemployment (Stevenson, 1977). Something of the fall in purchasing power in the district can be calculated from the records of the co-operative store presented in
Table 11.2. The figures indicate the pattern of the depression.
Between 1930 and 1933 profits, sales, average purchases and
credit all decrease with sales falling to just over half the 1921
figure. The fall in credit reflects both store policy and belt-
tightening among families.

Table 11.2   Profit, sales and credit, Throckley and District
             Co-operative Society, 1926-36

	Profit index (1921-100)	Sales index (1921-100)	Annual average purchase per member £	Annual average credit per member £
1926	86.9	56.5	42	3.3
1927	92.1	59.6	42	3.5
1928	93.0	59.9	41	2.9
1929	93.8	62.2	42	2.0
1930	94.4	62.9	41	1.5
1931	90.6	59.8	38	1.4
1932	82.6	56.2	35	1.3
1933	82.4	56.9	35	1.3
1934	92.9	58.1	36	1.3
1935	91.0	58.3	35	1.4
1936	87.5	59.2	36	1.5

Source: Adapted from Throckley District Co-operative Society,
         Tyne-Wear Archives

There is no doubt, even though it might not be precisely
measured, that many families in Throckley lived in poverty. But
there is no necessary connection between poverty and despair
or, for that matter, between poverty and protest. The critical
question is how people felt. I asked my aunt Eva whether they
felt themselves at this time to be poor. 'Wey no!' she said, and
went on:

Everybody thought the likes of us were well off! People
thought the Browns were wealthy. Me mother, for all that
she was hard up, had a proud nature. She was inclined like
that; and so was Maggie. Me granny used to say that 'Jim's
bairns are a bit proud, a bit stuck up'.

My grandmother was so far from seeing herself as poor that she
used to give away her hooky mats to those she thought needed
them, for each year, as I explained, she made new mats for herself. And it was her pride, Eva explained, which prevented her
from ever suggesting to anyone that she might be hard up. 'We
didn't shout stinking fish', is how Eva put it. They were not
angry at their poverty either. My grandfather was generally an
optimist. He used to tell my grandmother, 'Things'll be all right;
things'll brighten up.'
I put the same question to my mother. She said there was no
'real poverty' in Throckley, meaning by that malnutrition. They
were poor, she said, but they accepted it and got the best out
of life in other ways. 'And nearly everybody had gardens. They
might have had bacon bones instead of ham but nobody wanted
for vegetables.'

POLITICS OF DEPRESSION

Political activity has its roots, ultimately, in the way men experi-
ence the everyday problems of their lives. In Throckley, local
Labour Party politics acquired a distinctly radical tone during
these years and local political activity was closely connected with
the miners' lodges.
The two Throckley lodges were affiliated to the Newburn and
District Local Labour Party and through that to the Wansbeck
Divisional Labour Party. In the years of the Depression both
organisations were much concerned with unemployment and the
Means Test. By the mid 1930s with a deteriorating international
situation questions of peace and war began to take prominence.
The records of the Labour Party for this period reveal a well-
supported group of activists who, while not given to much
socialist theorising, none the less had a strong political awareness. They passed radical resolutions on the House of Lords,
on nationalisation and, in 1931, against those Labour cabinet
ministers who joined the National Government.
From the records of the Newburn Labour Party the membership figures for the early 1930s can be calculated in Table 11.3.
In comparison with present-day membership figures these are
very high indeed. Membership dues were collected weekly and
the annual general meetings of the party were so well attended
that the minutes had to be specially printed.

Table 11.3 Membership, Newburn Local Labour Party, 1931-6

	Trade Union members	Men	Women	Total individual members
1931	2,081	105	127	232
1932	–	116	142	258
1933	–	255	171	426
1934	–	266	189	455
1935	–	435	283	718
1936	–	337	256	593

Source: Local Labour Party, Minutes, NRO 527/B/1.

The activists in the party were convinced that the Labour Party was an integral part of the trade union movement and that industrial questions could not be separated from politics. Councillor Mavin of North Walbottle put this point nicely as early as 1927 when he reflected on the defeat of the lock-out (he noted in passing that he had lost his job through his political activity):

> The trade union movement and the Labour Political movement could not be separated. It [the defeat] further showed that if we are defeated industrially we must capture the political machine. The co-operative movement was also a very necessary part of the working class movement.... (Minutes, 1 February, 1927)

Despite the defeat of the miners and the parlous state of the industry after 1926, the policy of the local Labour Party on the coal industry was clear. In 1928 they resolved that

> the only way in which the Mining Industry can be properly carried out will be on the basis of Nationalisation with no Compensation to the present owners.... We pledge ourselves to work for that object. (Minutes, 10 March 1928)

In March 1930 they resolved to urge the Parliamentary Labour Party to abolish the House of Lords as they considered it 'against all the principles of Democratic Government for such a body to continue in office'.

The General Election in 1929 brought a second minority Labour government to office. It became clear early on, however, that the scope of this government's action was severely circumscribed by the economic crisis and orthodox financial policies. The Coal Mines Act of 1930 was one of its achievements. Its plan to reduce unemployment through public works schemes was widely welcomed. The Unemployment Insurance Act of 1930, which widened the scope of entitlement to a larger number of workers and extended transitional payment to those who had exhausted their rights under the scheme, was also a strongly supported measure. As

late as January 1931 the Throckley Labour Party reaffirmed 'its
faith in the Labour Government' and urged it 'to develop the
Home Market by increasing the purchasing power of the Masses,
whether employed or unemployed' (Newburn and District Local
Labour Party, Minutes, 1931).

Such Keynesian logic found little support in the Treasury, and
the financial crisis of 1931 destroyed the second Labour government. The story is well known; the version which is important
here is that many felt that the unemployed were being sacrificed
and the whole labour movement betrayed (Miliband, 1973;
Stevenson and Cook, 1977).

At its September meeting in 1931 the Newburn Labour Party
discussed the break-up of the Labour government. George
Shield, the MP, gave an outline of the events and Eddy Dowling
of Westerhope moved the following resolution: 'That this conference ... deplores the Government's refusal to re-consider
the cuts in unemployment benefit.' A marginally unsuccessful
amendment from the Throckley party was much more forceful:

> That this Conference demands the expulsion of MacDonald,
> Snowden and Thomas and urges the withdrawal of all Labour
> members of Parliament from the House of Commons for the
> purpose of rousing the Unemployed and Employed. (Minutes,
> September 1931)

My mother told me that many of the Labour supporters in
Throckley - including my grandfather - regarded MacDonald as
a traitor and many of those who had proudly displayed photographs of the Labour leader on their mantelpieces took them
down and destroyed them. Taking his cue from his friends, my
grandfather used to call MacDonald 'Ramsay MacBaldwin' and
'a bloody traitor'. At this time my mother's job involved collecting
weekly payments for the draper she now worked for, and at all
her calls she heard the same complaints about the Labour leaders
who joined the National Government.

In the general election of October 1931 George Shield the Labour
man lost to Lieutenant-Colonel Cruddas, a National Government
candidate in the Wansbeck division. In its review of the events
the Wansbeck Labour Party passed resolutions to elect future
Labour cabinets from the Parliamentary Labour Party, to create
a state medical service and, urged by the Throckley party, to
fight against the Means Test. The logic of this last resolution
was distinctly Keynesian. It was explained that since transitional
payments were being decreased more people were having to apply
direct to the Poor Law for relief.

> This in turn will mean a sharp rise in the Poor Rate, which
> is bound to reflect itself in increased rents for houses and
> add to the burden of the already harassed Working Classes,
> who now find it difficult to secure the bare necessities of
> life. (Wansbeck Divisional Labour Party, Minutes, October

1931, NRO 527/B/1)

COUNCIL HOUSES

If the national political scene was bleak, with the Labour Party
in disarray having suffered a massive defeat, the local opportunities for change were not so limited. The issue which illustrates this is housing. During the early 1930s the Labour council
pressed ahead with the modernisation of middens and, when they
could, council house building.

Housing problems remained severe in the district during this
period. As late as 1927 there were still over 2,000 privy middens
in the district. In 1930 the sanitary inspector revealed problems
of overcrowding even in the new council houses:

> The present cost of house building is a serious difficulty in
> a district like this suffering from the effects of prolonged
> trade depression. The present rate of wages of the working
> man does not admit of him paying a high rent and this has
> resulted in many of the council houses being sub-let, and in
> a number of cases even containing three families. There is a
> great demand for more and cheaper houses. (Annual Report,
> 1930; filed with MOH Report, West Denton Library)

Table 11.4 House building, Newburn Urban District, 1914-37

	Council houses	Private houses	Cumulative total
1914-1925	446	85	531
1926	39	9	579
1927	73	24	676
1928	12	7	695
1929	76	27	798
1930	20	19	837
1931	20	19	876
1932	–	34	910
1933	–	85	995
1934	225	148	1,368
1935	240	112	1,720
1936	56	111	1,887
1937	86	101	2,074

Source: calculated from the MOH Reports, Newburn UDC,
          Tyne-Wear Archive.

Housing had always been a major focus of political activity in
Throckley and throughout this period efforts were made to
improve its quality. During the 1920s some council houses were
sold in the district, but the policy of doing so was not considered successful, and during the 1930s under the provisions

of several Housing Acts the council pressed ahead with its policy
which Dick Browell had described in 1927 as 'plodding on trying
to build houses'. The problem was, he said, that 'the right kind
of house for the workers was required at a rent which the worker
could pay' (Newburn and District Labour Party, Minutes,
NRO 527/B/1). The building of houses was a central component
of their notions of progress, although at the height of the
Depression they were unable to build any at all. The figures for
the house-building programme are shown in Table 11.4.

The letting policy was to give priority to those in overcrowded
conditions, followed by those in houses with high rents and
those under notice to quit colliery property (Newburn UDC,
Housing Committee Minutes, Tyne-Wear Archive).

As the number of available council houses increased the turnover of families in the colliery houses increased, too; a growing
number of people preferred the better and more secure property
to a colliery house even though they had to pay rent for it.
Table 11.5 indicates something of the turnover in the street my
grandparents lived in. Mount Pleasant had something of its own
identity; it was almost a community within a community. However,
these crude and circumstantial data illustrate a change; the
community of the pit rows was not a stable one from the mid
1920s onwards. The aspiration of many people, particularly the
younger ones seeking homes of their own, was to get out of
colliery housing into something better.

Table 11.5 Population turnover, Mount Pleasant, 1914-36

% of residents	1914-20	1920-36	1914-36
Staying	84.7	30.2	28.0
Moving	15.2	69.7	71.9

Source: Newburn UDC, Rate Books, Tyne-Wear Archive.

This particular aspiration revealed itself in the Brown family
in heated discussions between Bill and his father. My grandfather, reflecting the generation he came from, believed in colliery housing. Bill, resentful that as a young miner living in
his parents' colliery house he was not entitled to a rent allowance on his wages, believed such housing to be a burden. In
any case, he argued, a colliery house kept a man tied to the pit.
Bill told me he used 'to gan hammer and tongs at me father
aboot that' and that he tried often enough through the union
to pass resolutions against colliery housing. Bill at this point,
i.e. the late 1920s, was anticipating marriage; he was courting
and his home was congested. The future seemed to him to be
very much in favour of his side of the argument.

FAMILY LIFE

The period covered in this chapter is a very significant one for
the family life of my grandparents. In 1928 my great-grandmother died at Eltringham and old Norfolk John was brought to
Throckley to live with his son Bob. Norfolk John lived at the
top end of Mount Pleasant so he saw more of his grandchildren
than before. But he died in 1931, his death causing some acrimony among his sons from which their friendship did not really
recover. The details of their arguments are no longer clear but
they concerned the distribution of bits of personal property,
funeral arrangements and the irresolvable question of who had
looked after the old man the most.

The most severe blow was, however, the death of aunt Maggie.
Harry, her husband, had retired from the pit in 1929 and had
to vacate his colliery house. They found a small cottage in Walbottle, disposed of most of their furniture and settled down to
retirement on a pension of ten shillings a week and four shillings
to pay in rent. After two years, during which time they were
helped financially by Francy, their daughter, and by my grandparents, Harry died. Maggie was not too well so my grandmother
did her washing and helped her occasionally with the house. The
main help, though, was to send my mother, now herself working
in a draper's shop in Newburn, to live with Maggie and look
after her. My mother agreed willingly to this. She had stayed a
lot with Maggie as a child and she wanted to get away from the
growing congestion of Mount Pleasant. The arrangement lasted
almost two years, until Maggie became ill and had to be brought
to Mount Pleasant, and that is where she died. My mother
remembers that my grandmother often said that she could 'never
ever repay the kindness my aunt Maggie had given us'. Her
death marked the end of an era for them all. For her nephews
and nieces her memory evoked warm images of a cosy childhood,
while the future in comparison looked bleak. My mother told me
how she felt when she came home: 'I hated it more than ever.
Our Eva and her kids were there, no private talks, always
noise of some kind.' But they all felt a deep sense of loss. Eva
said that Maggie 'was another mother to us'.

But these years were not all black ones for the family. Jim had
married in 1926 and, after a short period living in rooms in a
new council house, had secured a colliery house in Mount Pleasant, the one his aunt Maggie had lived in. In 1927 Jim's wife,
Nelly, gave birth to Carrie. My grandfather got on very well
with Jim and later liked to take Jim's second child - 'young
Jimmy' - to the gardens. Sundays were family days and in good
weather they all went for walks together in the afternoon.

Bill married a girl from Hexham in 1933 two years after he had
left the pit and, after 'living in' with his brother Jim for a while,
he was able to move into a council house at Throckley. By this
time, Olive, the eldest daughter, had moved to Newburn where
Tommy had a job as a gravedigger. Now the old people were

surrounded by their children and grandchildren. They liked
that, particularly my grandmother.

Even so, in 1931 there were still eight people living in the
house. The twins were both working and bringing in wages.
Eva helped in the house and looked after her two daughters,
Olive and Sadie, her marriage having completely collapsed. The
size of this household made my grandfather's role a little ambig-
uous. He was still the main breadwinner, father and grandfather simultaneously. He used to say, for example, of Eva's
two children that he was their father: 'I did everything but get
them.' But it worked. He carried on with his gardens and
animals. At weekends he went to the club. My grandmother
carried on, too, with the relentless routines of domestic work.
Louie and Jack spent as much time as they could away from the
house. They often went dancing at chapel 'socials'.

The really significant event of 1932, however, was an unofficial adoption; my grandparents took in the illegitimate child
of my grandmother's niece, Bessy. Bessy, working as a domestic
in Scarborough, became pregnant. Uncertain as to what to do
with the baby, and considering having her adopted, she wrote
to Eva from the Poor Law hospital in which she was staying, for
advice. Eva discussed the matter with her mother and it was
agreed that Bessy should come to Mount Pleasant with her
daughter, Gloria, and stay with the family while she decided
what she was going to do. She came to Throckley and stayed a
few weeks. Having the baby adopted was discussed as a possibility but my grandmother solved the problem by agreeing to
take Gloria herself. Bessy agreed and left, travelling to Bolton
to seek domestic work. At the age of 61 my grandmother acquired
a new baby, the second time in her life she had taken her relatives' children into her own home.

'Gloria', my mother told me, 'gave them a new lease of life and
kept them young.' Gloria was fed on milk from a goat my grandfather had at the time. She was a kind of doll to Eva's two girls,
and she grew up to think of them as sisters. Bessy made intermittent contact with her daughter, but although Gloria knew
from when she was old enough to appreciate it that her 'granny
and granda' were not her parents she thought of them as such.

Gloria was not, in fact, a great burden to the family; there
were several Brown grandchildren living near and there was
plenty of immediate help in Mount Pleasant itself. Kinship, then,
was what made the unofficial adoption of Gloria possible and her
presence, I suspect, contributed in its own way to strengthening
the central position of Mount Pleasant in the family life of all
the Browns.

During this period some of the strains of congestion did surface, however, and tempers were strained. The presence of
small children raised questions of who was getting more than her
fair share of attention or who was being treated too leniently.
Jack and Jouie felt only marginally connected to the house at this
point. Jack could rise quickly to anger and felt that he had

little privacy. He did not interest himself in the gardens –
indeed, he was not encouraged to do so; my grandfather used
to tell him, 'The best side of the garden for you is on the other
side of the fence' – so he could not beat the same escape as my
grandfather. He had to look for his relaxation outside the
house. In the early 1930s he was unemployed for a while and
this made him feel outside of things.

Unemployment was also a source of conflict with my mother.
To earn a bit of money Jack set up a small shop in the garden
shed selling home-made ginger beer, some vegetables from the
garden, sweets and cigarettes. The cigarettes he bought from
the co-op and through that increased his co-op dividends. He
also sold small items of drapery on a payment plan system. It
was this which brought him into conflict with my mother because
her job at the time was to do precisely the same thing for the
draper she worked for. She thought Jack was taking work away
from her and they argued about it.

None of these conflicts was incapable of solution, and when
my relatives recall these difficult years they invariably refer to
the great integrative rituals of the weekend sing-songs and
Sunday evening suppers. It is difficult for them to re-live the
disputes; to do so might even open up feelings of acrimony
which no mechanism now exists to control. Sufficient to note,
therefore, that their family life was not always harmonious; it
was lived under conditions of great stress the source of which
was overcrowding.

RETIREMENT

Outside the family my grandfather was still preoccupied with
the pit. Still in his early 60s, he knew he could have up to ten
years of working life yet, time enough to get pulled around and
keep a secure home for them all. Pit work carried the same risks
as it always had, however, and from this time on he had another
worry. He was starting to get 'dizzy turns' in the pit. His heart
would quicken to a very rapid pulse and he would feel light-
headed. The root cause of this tachycardia was never established
but the doctor advised him to leave the pit. He did not accept
the advice at first and carried on. In every other respect he felt
strong enough to do so and able to live a perfectly normal life
and to relax more than in the past. In 1934, for instance, he
went on a rare bus trip to Scarborough with the club. But it
was also in 1934 that he had another pit accident. A falling stone
from the roof glanced past his head, nearly severing his nose
from his face. Under the convalescence scheme available through
the club, he spent a few weeks at Saltburn Home recovering
from the blow and, he hoped, from the dizziness which affected
him underground. When he returned to the pit he was put on
lighter surface work. This did not suit him because the pay, at
6s. 9d. per day was low. He always tried to work on Sundays

to get an extra day's pay. The dizziness persisted, however, and his doctor told him emphatically that if he continued to work he would be dead within a couple of years; without working he could go on till ripe old age. At the age of 63 he conceded the point and left the pit. He had no pension although he did have £1 per fortnight from Heddon Club, the Friendly Society.

He had worked for fifty-two years, thirty-five of them for the Throckley coal company. There was no dramatic finish to his career. He had been ill and off work and he simply never returned. He used to say; 'I never worked a notice and I never got a notice.' He did not even bother to retrieve some of his pit gear which he had left underground. There was no collection for him, no gratuity, and a rent, as they say, was 'put on him' immediately, though he and my grandmother were allowed to continue living in the colliery house. My grandmother used to say that they would be a bit better off when they got their pension.

He had no regrets about leaving the pit. He said often enough that men should not have to work underground and that they would be better off when all the pits were closed. However, such uncharacteristic equanimity was rooted in illness. He was so ill that my grandmother did not think he would see the Christmas out. In anticipation of an untimely death she arranged a great family party not, of course, to celebrate, but to cheer him up a bit. He recovered, though, and settled into a strict routine of visits to the garden, and relied on his sons; one of them, Jim, being at this time a committee member at the working men's club, would take him out at weekends for a drink.

He kept up this pattern of life for practically another thirty years. Illness always threatened him; he did take things a lot more easily. But my grandmother always reckoned he perked up at weekends. 'Thou's always alreet at the weekend, Jim. Poorly all week till the pubs are open!' is what she used to say. The club and the garden kept him in contact with his friends. Retirement for him was not a great rupture to routine; after he got over his ill-health, it was, at least in the early years, a period of relaxed and purposeful freedom.

12 RETIREMENT, WAR AND RIPE OLD AGE

The first few years of my grandfather's retirement were built around strict routines. In contrast to his working life when he got out of bed at irregular hours he now got up punctually at nine o'clock. Eva told me that his feet would touch the floor at exactly the same time as the grandfather clock chimed the last stroke of nine. He moved about the house slowly, 'That owld bugga's too slow to catch cold' is what his wife used to say of him. Eva claims that they could practically set their clocks in Mount Pleasant by watching the movements of my grandfather between his gardens and his home. Bill said of him:

> He was a contented man. He never grumbled. He was like that all his life. He accepted retirement. If he wasn't at the garden he was sitting on that seat in front of the house. Mebbe he would get up to pluck a weed. He waved to people who went by. The old men used to hang around the road ends. But me father wouldn't go.
> If there was a cat come, ten to one he'd hit the bugger. He was just content.

He was not isolated, however. He had three small children in his house and grandchildren nearby. He had many friends who met in the gardens and in the club. He took an active interest in his family and family gossip.

And it has to be noted that the family itself – and at this point I write as my own respondent – is an extremely introverted one. Physical proximity meant that they could maintain contact with one another and swap an endless flow of what would have looked from the outside like trivial gossip. And talk! My relatives do not talk to one another; they shout. Very often one does not listen to what another is saying. Some of my earliest memories are of congested evenings in Mount Pleasant. Through a fog of smoke in the odd, bright glow of the gas mantle, and across the red-tasselled table, I can clearly picture the people. I can see the bending, stooping movement in and out of the kitchen, and there is the overwhelming impression of talk. Everything was reported. There was no filtering process for irrelevant detail. It really did not matter whether anyone was listening; to be heard at all was a miracle and to understand it required an inside knowledge of the family and its feuds which no stranger could ever possess. I am convinced that my grandfather led such a routine retirement as a subtle form of escape from the family itself. Routines gave order to what otherwise might have been chaos; he found in his retirement what my grandmother had understood for nearly forty years.

His own life may have had predictable routines which little in the future might conceivably change, but these were hectic years for his own children. Olive, Jim and Bill had left home; Eva had returned. From 1936 onwards Louie was courting a lad from Denton Burn, the son of a miner but a factory worker himself. Jack had several girlfriends and was likely to leave home at any time. By 1938 he had a good job on the railway having previously worked for a firm of undertakers in Newcastle (his second such job).

Young Olive was a teenager by the time my grandfather retired. He took a stern interest in her behaviour ensuring, when he could, that she came in at nights at a reasonable time, and that she didn't play the dance music, of which he disapproved, on the gramophone too often.

Louie, still working at the draper's shop, married in 1938 and took rooms in a new council house opposite Mount Pleasant. In April 1939 she gave birth to a son, Jack, named after her brother, and shortly afterwards moved to her own council house in Scotswood.

Jack, courting a girl from Gateshead, married shortly afterwards, and after one year in Lemington he got his own council house in Throckley.

The fortunes of his family were the framework of my grandfather's own life. The gossip, the problems, the visits and the bairns were what, apart from the garden, dominated the years immediately before the war.

THE APPROACH OF WAR

International news did not preoccupy my grandfather. It was not that he was uninterested. Rather, he felt that matters of international politics were remote from the likes of him. He did discuss political matters occasionally. Bill remembers him in a worried discussion with a friend, George Gregory, asking, 'What's going to happen with this Hitler, George? The German re-armament's not good.' And when Chamberlain came back from Munich with his piece of paper promising 'peace in our time' my grandfather was not in the least convinced. He once explained to Eva that Chamberlain - whom he described as 'an aa'd knave' - would not mind war and that 'ther's aalways plenty of work when the Tories are in: work for ammunitions to kill folk'. In his opinion the Tories were warmongers. The Northumberland Miners' Association had always, of course, maintained a pacifist stance, and as early as 1933 raised an alarm about Hitler and Fascism. On 20 May the council of the NMA passed the following resolution:

That this meeting hereby enters, in the name of democracy,
its strong protest against Hitler's crushing of trades unionism
in Germany, and the atrocious treatment of old and experienced
trades union leaders in that country. Further, we urge the
British Government to take the necessary steps in the name
of humanity, to show Germany its resentment against the
atrocities committed by gangsters acting with the approval of
the Hitler regime.

In his October circular of 1933 Bill Straker warned his members:
'Hitler promised liberty to the German people and gave them
slavery of the worst kind.'

The Spanish Civil War became an important and emotional
question for Labour activists in Throckley. Nobody from Throckley joined the International Brigade, but the Newburn and District Labour Party declared itself in October 1936 to be 'profoundly alarmed at the present state of affairs in Spain' and
urged the Labour Party itself to change its neutralist stance of
non-intervention. In January 1937 the party joined in a campaign
to sell tickets for the Spanish workers' medical relief fund and
organised street collections. The co-operative store was also
involved in this fund raising, encouraging people to donate their
dividend, and in one campaign persuading people to buy milk
tokens which would then be passed to the Co-operative Union
to buy powdered milk for 'democratic Spain'.

By 1938 the League of Nations, the institution which, after
the First World War, had been seen by many to be a guarantor
of peace, had obviously failed. Recognising that, Sir Charles
Trevelyan Bart proposed the following successful resolution to
the Wansbeck Labour Party:

That this conference ... considers that the unchecked aggres-
sion of the Fascist nations is directly due to the weakness,
the Fascist sympathies, and the criminal acquiescence of the
present British Government and that there is no security for
the saving of democracy unless the Chamberlain government
is changed at once, and replaced by a new government which
believes in an alliance, within a recovered League of Nations,
of Britain, France and Soviet Russia, with the support of the
United States and the remaining democratic countries, and
calls on the Executive Committee of the Labour Party to sum-
mon an emergency conference to deal with the greatest crisis
since August 1914. (Wansbeck Divisional Labour Party, Minutes,
March 1938, NRO 527/B/1)

In 1938 the Newburn Labour Party organised a May Day demon-
stration which was addressed by Aneurin Bevan, who was then
campaigning vigorously for a Popular Front, and some local
speakers, together with 'a wounded member of the International
Brigade'. Trevelyan Bart and Bevan were subsequently expelled
from the Labour Party for their Popular Front views.

In March 1939, at the annual meeting of the Wansbeck Labour
Party the prospective MP, Mr McLean, attacked the policy of
appeasement of the Chamberlain government.

> The Chamberlain pilgrimage of appeasement has resulted in
> betrayals, mad race of armaments, and the strengthening of
> the dictator countries. The greatest betrayal of all is our
> own dead, which will ultimately lead the living to destruction.

At the same meeting two of Newburn ward's resolutions were
carried, one calling for a common programme to fight Fascism –
showing them to be greatly in sympathy with Bevan's position –
the other calling for the nationalisation of the armaments industry.

WAR IN THROCKLEY

Paradoxically, it seems to me at least, my relatives find it difficult to recall much that is outstanding about the war. Their
lives were highly routinised and little different from the period
before the war. Throckley, of course, was affected hardly at
all by bombing, and the Brown family had no one in the forces.
At the beginning, however, their dominant mood was one of
uncertainty.

The outbreak of war in September 1939, following Hitler's
invasion of Poland, was expected but was nevertheless a shock.
Jim's wife, Nelly, told me: 'It was Sunday morning. The sirens
went at 11 o'clock. We got a shock. We didn't know what to
think.' What they expected was aerial bombing and perhaps
gas attacks, gas masks having been issued. Bill was able to tell
them more about arrangements for dealing with the casualties
for he had been an active member of the St John's Ambulance
Brigade for a few years. The first few months were, of course,
very quiet, but there was plenty of evidence of mobilisation
which affected the Browns directly.

In September 1939 there was a march-past up Newburn Road
of the 5th battalion of the Royal Northumberland Fusiliers. My
grandfather watched this from the end of his garden. Olive's
boyfriend, Jack Danskin, was in the parade and they were eager
to see him march by. My mother remembers that as they stood
there my grandfather said how it took him right back and he was
thinking of 'wor Gordon', the lad he had brought up and who
had been killed early in the First World War. And he turned
with tears in his eyes to my mother who was holding her baby,
and said to the baby, although it was a comment directed to
them all, 'Son, you'll finish this lot off.'

My grandfather expected a protracted war, a repeat of 1914.
But he was not confident that Germany would be defeated. Like
most other people he was simply anxious and uncertain.

The battery powered radio was a major source of information

about the war as were the newspapers. In addition Olive took
'War Illustrated', the weekly picture magazine edited by Sir John
Hammerton. They all read this thoroughly, taking a keen interest
in military affairs, an interest fuelled in Olive's case by her
fiancé's censored letters from the army.

Like every other family the war affected the Browns directly
if not dramatically. Jack was turned down for military service
and classified as grade five fit on account of a damaged arm.
Jim, at that time working again at Spencers, having left the pit
on account of its poor pay – a shift, incidentally, indicative of
the changes which had already occurred in the character of the
community in Throckley – was declared to be working in a reserved occupation. Jim amplified for me his reasons for leaving the
pit, indicating, perhaps, that the halting attempts of the
Throckley coal company to modernise and to instal coal-cutting
machinery were actually driving some of the men from the pits
by devaluing skills and ruining the work environment.

> I left cos the money was bad. One of my marras left to the
> Police. They were aall trying to get oot. There was no pro-
> spect in the pit. It was bad work; it was rough. With the
> coal companies' machines everything was a rush.

Bill was on full-time ambulance and ARP work. Young Olive was
directed to work in munitions for a short period in the south of
England, but returned to munitions work in Vickers Armstrong
on the Tyne, milling Merlin engines for bombers.

An Anderson shelter was installed in the front garden in case
of air raids, although my grandfather was reluctant to use it.
They put the blackout curtains up but he preferred to walk up
and down the garden watching the night sky to sitting in the
shelter.

In the whole course of the war there were 298 bombing raids
over the northern (no. 1) region covering Tyneside, Northumberland, Durham and the North Riding of Yorkshire (Lambert,
1945). Fatalities amounted to 1,447, with more than 5,000 people
injured. In the Newburn Urban District only three people were
killed and twenty injured. These figures, in comparison with
those for the large cities in the Midlands and south-east, are,
of course, slight. The regional commissioner attributed the
failure of the Luftwaffe to destroy Tyneside to three factors,
the quality of the defences, the geography of the rivers in the
region making them poor guides for aircraft and, finally, 'the
natural industrial haze constantly obscuring the vital targets'
(Lambert, 1945, p. 7).

One bomb did drop in Throckley Dene causing extensive blast
damage. 'What a bloody noise!' said uncle Jim. It slammed his
doors, broke windows. Young Jack hid in a cupboard under the
stairs. It brought the ceiling down in uncle Jack's house and
really scared my grandfather. When he heard the blast he dived
into the Anderson shelter shouting, 'Where's me hat?' My

grandmother responded to this with, 'Never mind your bloody
hat as long as your bloody head's on.' 'The next morning,'
Sadie told me, 'we all went looking at this big crater. The kids
were all looking for bits of aeroplane glass and shrapnel to make
rings and bracelets. All the greenhouses were shattered.' Had
the bomb landed on the housing estate nearby the casualties
would have been serious.

Apart from this episode the only events which brought the
war in the air close to them was a loose bomb near Walbottle and
an aircraft being brought down by a barrage balloon between
Walbottle and Westerhope. Uncle Jack was required to billet some
airmen during the war since they had some spare accommodation.

The war directly affected working hours and life-styles; long
shifts, ARP work, rationing, blackouts and bombing raids forced
families into highly predictable routines. And all families have
their stories of bizarre events during the war. My uncle Jack,
a railway worker, was nearly killed near Hexham when an
ammunition dump near which he was working exploded. He was
also once taken for a spy by a schoolmaster at a country station
and was locked up in the schoolroom till the police came. After
that all railway workers were required to carry identity cards.

In the main, however, the war was not bizarre; normal life
went on but under increasingly difficult conditions. Some of
these difficulties come through in the half-yearly statements of
the co-operative store. In April 1940 the store committee noted:

> The incidence of the Great War is felt in every village. Our
> kindred stands on foreign soil, between us and the enslave-
> ment that would come with defeat. We at home will be called
> upon to sacrifice such fancies as we have indulged in and eat
> many kinds of things that in peace time we would refuse. We
> can offer for sale only the meat the Government sends us....
> We shall not starve if we conserve our resources. Keep the
> home front steady and our shopping dignified. (Throckley
> District Co-operative Society, Balance Sheets, Tyne-Wear
> Archive)

And in January 1941 the committee returned to the same theme:

> We regret the severe restrictions now operating in this depart-
> ment, but we are now enduring a ration much less than our
> customary sales per member. It is the price we are now being
> called upon to pay, that a speedy victory may be ours. We
> are fortunate that our supplies are forthcoming unaccompanied
> by a blitz and we ought cheerfully to accept no more than
> the prescribed ration ... and we urge our members to accept
> the situation with what cheerfulness and tolerance they can
> muster until victory is secured.

Sadie told me that the main shortage in Mount Pleasant was butter.
They did not get any bacon ration because they kept their own

pigs, and the gardens kept them well supplied. Jim, Jack and
Bill had gardens as well as my grandfather and they shared
among themselves. There was a shortage of coal, too. The supplies of domestic coal were strictly controlled, although not
rationed, from January 1942 onwards (Court, 1951). Miners still
received their allowance, but as a retired miner my grandfather
got less than previously and invariably had to buy in extra.

Typically, however, the self-reliance theme which dominated
his life reasserted itself. He began to collect coal regularly from
the dilly line or from the streets themselves if a load had not
been properly 'shovelled in'. Gloria told me that if they, the
children, were playing and noticed that coal had fallen off the
dilly onto the line they always ran home to tell their grandfather,
and he would go straight down and get his bucket. On occasions
the men on the dilly 'accidentally on purpose' would make sure
some coal did fall off for him to collect. And during the wartime
summers he used the method of mixing coal dust with soapy water
to make 'duff balls' to supplement supplies, a technique, he
wryly commented, which they discovered in 1926 and which the
government was now recommending to conserve coal.

The war emerges mostly in their memories as a matter of routine. There were occasional house parties, where bingo was
played, up and down Mount Pleasant to raise money to help
soldiers. Many of the women in the street, including Eva and
my grandmother, knitted socks and squares for blankets. They
listened to the radio, and complained at Lord Haw-Haw despite
Nelly's theory that he really was on the allied side telling them
exactly where bombs would be dropped. My grandfather particularly liked Workers' Playtime.

The political management of the war seems not to have bothered
him much. He was suspicious of Churchill and wondered after
he became prime minister in 1940 whom he might turn the guns
on. Jim's view is perhaps similar to his father's with respect to
Churchill as a war leader. Jim told me, 'Attlee and Churchill
worked well in the coalition although Churchill got the praise.'
To the miners, at least those in my grandfather's circle of friends,
Churchill was a man beyond forgiveness.

What my grandfather attended to with most seriousness during
the war was his family. And the main concern was the children.
Gloria and Eva's daughter Sadie were still young. He had nine
grandchildren living close to him by 1939. In December of that
year Jack's wife, Mary, gave birth to a boy but the child did
not survive. The death of babies was, then, something which
still haunted them; Jim's wife, Nelly, had also lost a baby in the
late 1920s. Both deaths were upsetting but both reflect a much
larger problem. Infant mortality on Tyneside in 1937 was 82 per
thousand births, an improvement on the 1925 figure of 93, but
still much worse than the figure for England and Wales as a
whole which was 58 (Goodfellow, 1941). Children were, however,
a source of much pleasure; each was indulged. During this
period my grandfather kept a few geese and used the goose

down to make quilts and layettes for the babies. Each child
activated the old routines of christening parties. The older ones
were taken for walks, or taken to the garden.

Gloria and Sadie both remember long walks with my grandfather. He took them down by the river with Bill's dog and
walked all the way back through Heddon, a route he had tramped
often enough with his own children. He clearly enjoyed these
walks and, though a normally quiet man, used them to talk about
the past. Gloria still has vivid images of the terrible stormy
night when, for the one and only time in his life, my grandfather,
as a young man, turned back from the pit, frightened to work
through the night. He told them about his horses, his work as
a lad, his days in Heddon. He showed them how to trap rats,
to help chickens to hatch, to steal a turnip. And he told them,
too, about 1926, conveying to those girls his own values and
those essential elements of social awareness which were part of
mining and its past. The images were not just of exploitation,
although he did tell Gloria how he had worked harder out of the
pit during the 1926 strike than ever he had in it and all the
Brown children and grandchildren will retail his dislike of
Churchill and the Tories. He communicated, too, a sense of a
good life of neighbourliness and co-operation and he never
changed his view that the Throckley coal company had been a
good one to work for. If they saw any of the Stephensons when
they were out on their walks he invariably raised his hat to
them.

The girls were clearly a vehicle for a kind of life review.
Gloria says that, looking back, it is clear her grandfather was
totally content and keenly interested in what went on around
him. He knew every ditch and tree and field and was totally
happy in the open air. Bill acquired a car in 1940 as part of his
growing insurance business, and he sometimes took his parents
out on runs into the countryside. On one such run he hit a
rabbit and the old man asked him to stop the car; he got out
and with his penknife gutted and skinned the animal - 'and that',
said Bill, 'was a meal'. He never missed a chance and remained
alert right through his long life.

POST-WAR SOCIAL CHANGE

The end of the war was celebrated with street parties and
dances. The dominant mood was one of relief. Eva told me that
there was singing in the streets and all the streets were decorated. 'It was lovely to think we could have our windows lit up
again.' However, there was no sense of a new world to be built.
'Not in this village', said Bill when I asked him about the high
expectations that some historians described. And Mary, Jack's
wife, was more emphatic: 'There was nothing to be optimistic
about; we were making do and mend, making coats out of blankets!' My mother says that she expected a major economic

depression after the war. The result of the First World War had
been depression; they did not think this one would be any
different.

The ending of the war was marred for my grandparents by the
deaths of two of their grandchildren. My twin brother died at
the age of 9 months of bronchial pneumonia and convulsions in
1945, and Jack and Mary's daughter died of the same illness at
the age of 18 months. In that same year young Olive gave birth
to her son, John Danskin, the first of the great-grandchildren.
Francy, Olive's daughter, who married a soldier, gave birth to
Ronnie Harvey. In both cases the marriage faltered quickly.

Politically, the country was poised for a period of massive
social change, and the exhausted coalition government gave way
after the general election of 1945 to a Labour government under
Attlee. This news excited my grandfather. Gloria remembers
sitting up with him through the night listening to the radio as
the results of the election came in. Alf Robens was elected for
the Wansbeck division, the first Labour MP for the division since
1931. The Newburn and District Labour Party worked hard for
this result and looked to the repeal of the 1927 Trade Disputes
Act as a first priority for a Labour government. In 1944, with
the end of the war in sight, the annual meeting of the Wansbeck
Labour Party passed the following resolution from the Newburn
ward indicating quite clearly that nothing less than a Labour
government afterwards would meet their demands:

> That this conference feels satisfied there is no hope of this
> Government making any fundamental change in our economic
> structure to provide a reasonable standard of life with
> security for the people and, feeling just as satisfied that the
> great majority of the people desire it, and are most certainly
> entitled to it, we, therefore, urge the Labour Party to give
> a bold and determined lead for the socialisation of Land,
> Finance, Industry and Transport, and a guaranteed National
> Minimum weekly wage, believing by so doing they will give
> new life and hope to the toiling masses in industry and the
> services and at the same time awaken an enthusiasm and a
> faith in the party that is so urgently needed. (Wansbeck
> Divisional Labour Party, Minutes, 1944, NRO 527/B/1)

The experience of war, the need to mobilise the whole civilian
population, the changes which were introduced to deal with production, casualties and labour mobilisation and the commitment
of both the main political parties to economic planning, had all
combined to produce an entirely new set of opportunities and
circumstances for political action. In the five years after the war
the framework of what came to be understood as the welfare
state was set up. Leaving aside whether these were socialist
measures or not - even whether they were effective - it is important to stress that for my grandfather's generation, they represented the highest of their political hopes. They brought the

nationalisation of the pits and the railways; the 1927 Trade
Disputes Act was repealed; they brought a universal social
security scheme and abolished, so it was thought, the Means
Test. The National Health Service was set up. Planning and full
employment were key themes in all government policy. And the
most obvious change was in council house building. Lamb's
fields just below Mount Pleasant were designated for housing
development, and from 1946 onwards a large, extremely pleasant
and well laid out estate was built, known locally as the White
City. Council housing again became the key theme of local politics
with a Labour council the 'natural party of local government'.

During this period my grandfather was in his mid 70s, alert,
but not too clear about precisely what was happening. Gloria
brought this out nicely in respect of the National Health Service:

> I don't think my granda understood the change. Gladys
> Dailtree came collecting for the Doctor's money and the Labour
> money. They had a card. If she didn't come Lizzy Storey
> came. When they changed and health stamps started he didn't
> click on to that. He didn't click on that you could go to the
> doctors and not pay. He hadn't paid for it so he didn't think
> you should have it. He didn't like getting two shillings off
> his prescription. If he had worked and paid for it ... that
> would have been a different matter.

But what she is picking up here, of course, is the lifelong theme
of independence and his utter dislike of charity. He did not feel
he had accumulated rights under the scheme and he therefore
questioned his own entitlement. When he was clear about his
rights he was prepared to insist on them. If the children could
not play at the Throckley Welfare ground he would say angrily,
'We paid for that; we're entitled to it.' It was the same after the
end of the war with his ration books. For a while the store had
a scheme to give Australian tinned jam to pensioners but he
wouldn't go to get it.

He did, however, try to keep himself informed about politics
and world affairs and he and Bill used to discuss politics quite
intensely. Alf Robens, the MP visited his house on several
occasions and stayed for meals. Nobody can remember what they
talked about but Gloria says of these visits: 'We had to be on
our best behaviour at the table. He was given the best.'

Nationalisation of the mines was for the miners a great achievement. The mines had effectively been under state control throughout the war, although not free of conflict on that account. The
period immediately after the war was difficult for the industry.
Reorganisation plans had to be drawn up in the face of labour
shortages and coal shortages. But it came. It came, Kirby says,
'with confusion and disillusionment in the minds of managers and
workers alike' (Kirby, 1977, p. 200). In Throckley the transfer
of the pits to the National Coal Board went smoothly, the Throckley coal company being compensated for assets assessed at

£236,328 in 1952. The company had already been paid £138,889
by the time this assessment was made. The management changed
more quickly. Jim told me, echoing the views of his father: 'Best
thing that happened when they nationalised the pit. The people
got a decent wage and everything then. [But] the cars started
to drive into the pit yard. They didn't know who they were.'

Old Tom Stobbart told me that nationalisation was a fine thing
but they did it the wrong way; some of the old gaffas were still
there. Uncle Jack complained about nationalisation on the railways. He used to tell Mary that they had more bosses than
workers. Where they always used to work with new wood, they
now got old wood, and the quality of work was worse. The only
places which got new wood were the offices of the bosses. It
does not matter whether facts are accurate; they were thought
to be accurate. My grandfather reserved his judgment on
nationalisation and he retained for a long time a distinctly romantic notion of it. In 1956, for example, the National Coal Board
modernised the houses at Mount Pleasant, installing a new fire-
range and electricity. My grandfather was worried about this
believing, as Eva told me, that 'the miners were trying to do over
much too quick'.

OLD AGE

James Brown was a remarkably busy man whose family life had
been unique. His own family had grown up by 1933, but from the
late 1920s onwards he took on his daughter's family. By 1950
with the break-up of young Olive's marriage with Jack Danskin
he took in his granddaughter and her child, the third time in
his life when he had taken in other people's children. As he was
involved as a surrogate father for two successive generations
it is difficult to say when he became old.

The turning point, I think, was 1948, the year of his golden
wedding aniversary. This was celebrated with a huge family
gathering in Mount Pleasant and it was a very proud day for
both of them.

By this time their worries, in fact, were few. It is clear in
retrospect that during this period those who were retired were
in considerable risk of poverty. Abel-Smith and Townsend (1965)
found in their re-analysis of data for the early 1950s that two-
thirds of those families with low expenditure had a retired head.
But because of the peculiar composition of the Brown household,
and the net addition to their real income from the gardens, they
were not poor. They were, in their own words, 'hard up' but not
poor. Gloria, Olive and Sadie were all working and contributing
to the household income. Eva helped Bill occasionally by collecting for his insurance business, and Bill himself was doing well.
My grandparents felt more secure in their old age than they had
done throughout their whole married life.

The old were indulged, at least the men were. The two clubs

in the village organised trips and gave annual treats to old
people. It often happened that all retired club members were
given a half-bottle of whisky for Christmas. The local authority
had placed several sheltered seats around the district where
the old men could sit and talk away the hours. My grandfather,
as Bill explained, did not mix in that way. He kept his socialising
for the weekend. In the Union Jack Club he had a special seat.
If it was occupied when he came that member would usually know
that he was expected to give it up. When this happened my
grandfather simply smiled and nodded and took up his position.

The most poignant recognition of the age status came, of
course, from within the family. The old people were a kind of
fulcrum around which the extended family revolved. Weekend
nights, New Year and Christmas Eve were important occasions
on which the family gathered together. From being a small child
I can remember those weekend visits to Mount Pleasant. The
whole scene seemed focused on my grandfather; people deferred
to his age, and just before he arrived back from the club with
the men the women hurried to put the final touches to the supper. Sadie and Olive hastened back from the dance to be there
on time. Sadie and Olive used to stand, their backs to the fire,
their skirts lifted up to get warmed. Some of us younger ones
used to see if we could catch a glimpse of their knickers. I'm
not sure whether we ever did but at least we tried. But when
the old man came in everyone moved away from the fire so
that he got a good open view of it. He was helped off with his
clothes and shoes and fussed over. This pattern of indulgence
went on till the day he died, increasing in intensity as he became
more and more frail.

From 1950 onwards my grandmother became frail. Circulatory
problems coupled with a weakening heart slowed her up; heavy
colds took their toll, too. She died in 1953, coronation year. It
was not unexpected. Since the death in the Maria pit of her
neighbour Jack Batey, killed by a fall of stone, she had been
depressed and poorly. My mother is convinced that Jack Batey's
death was the start of her own mother's decline. She never
really recovered from a dose of flu which for a while had left
her in a coma from which some of the family thought she would
never emerge. Eva says that Francy had told her firmly, 'Get
the dead clothes out.' She survived for a while, all the time in
her own bed in the alcove of the room.

She had never wished for an ostentatious funeral, simply
cremation. Her coffin stood on trestles alongside the piano in
the front room and life went on around it. One of the more vivid
memories of my childhood is being led to look into that coffin
and to see my mother kiss my grandmother for the last time.

The old man never really recovered from her death. He tried
to carry on with his gardens and he managed to do so for another
four or five years. He got through it, I think, by putting Eva
in the role of his wife. From this time on, however, he was often
confronted with death – being old himself he knew so many other

old people. I often thought that my grandfather's generation literally died on him.

Uncle Jim Stobart also died in 1953. Olive, my grandfather's eldest daughter who was by then living in Whitley Bay, died of cancer in 1958. That was a blow to him; he was deeply attached to her and said that he should have died instead. In 1963 Francy died. Several of his brothers or their wives had died, too, by this time. He attended many funerals in his old age, each one, I suspect, confronting him with the imminent possibility of his own. He discussed this with Eva, expressing many times a wish to be buried at Heddon-on-the-Wall.

TO A COUNCIL HOUSE

My grandfather was never on the periphery of his family life. During his last few years at Mount Pleasant he was directly involved in the lives of a grandchild and a great-grandchild. Young Jimmy Brown was struggling hard after his national service in Egypt to build up a smallholding. His grandfather helped him with the pigs and chickens. And John Danskin, young Olive's son who had lived with him since he was a child, had always been keen to raise chickens and keep animals of all kinds, rabbits, whippets, pigeons, goats and pigs. His great-grandfather helped him a lot, too.

In 1964, however, Olive remarried, and with Sadie about to be married there were only himself, Eva and John Danskin in the house. Eva was offered a council house. The old man was reluctant to move but Mount Pleasant was scheduled for demolition. His surviving children are agreed that the move killed him; it transposed him to an unfamiliar environment and lifted him out from a complex of memories and associations which he had woven into a personal image of a full, rich life. The council house had no garden to speak of. It had an inside toilet and he thought that was unhygienic. He told Eva that he thought she had a nice house, never conceding it was his, too.

He was now very frail. His daughters shaved him and washed him. Worried by incontinence, he was reluctant to go out for a drink, so he remained inside, always, of course, receiving regular visitors.

LIFE REVIEW AND SOCIAL CHANGE

I once asked my grandfather how he would have arranged his life if he could have lived it again. He was very old and living in Eva's council house when this conversation took place, but he was in full possession of his faculties. His reply was something like this: 'If I had the chance I would do just the same again. We had some hard times but we had some good times, too.' Two points arise from this. The first is the apparent irrationality of it. Knowing what he had lived through it seems inconceivable that he could choose to do it again. But I think that that is not what he meant, for he could be quite critical about most aspects of his life. In carrying out his 'life review' (see Thompson, 1978) he would 'naturally' consider his experiences selectively in such a way that he could rescue a credible and worthy biography. Society helps in this by providing conventional formulae governing how to talk about the past.

My grandfather was retailing to me a conventional cliché. It was, however, a cliché with great personal significance; I do not believe he was either lying or deceiving himself. When I asked him the question his answer was honest and authentic, although I think he was mildly surprised that anyone could ask such a stupid question; he was not a man who dwelt much in the past and nothing in his own experience - at least, not much - encouraged him to be 'reflectively aware' of his own biography. My point, then, is that old people draw on conventional ways of talking about the past when they review their own biographies.

The second point is this: in saying he would live it again in much the same way he was articulating the view that while much had been difficult in his life he could none the less see in it the forward march of progress. And while one 'reading' of the history of 'the common people' might dispute that much of any worth was gained at all - an argument which my uncle Bill used to taunt my grandfather with, especially when Bill was being critical of the Labour Party - it is undeniable that, contrasted with his early life before the First World War, his retirement and the world around him generally were incomparably better than anything he could have imagined. It might be ironic that having lived through some of the most difficult periods of the twentieth century he could look back with such a benign view of progress.

He measured such progress by the progress of his family. In each of them, in different ways, he saw the unfolding of social change. Of his three sons, two were totally unconnected with the pits; his family were decently housed and three of them, Bill, Olive and my mother, were running their own small businesses. Most of his grandchildren had been to secondary schools, some to the grammar school and higher education. All of them had jobs; some of them were running around in cars. They all had television. And I suspect he felt pleased that he himself was still well respected in Throckley and that, in the face of much that might have destroyed it, he had kept his self-respect.

In the last few years, however, he did not feel that well known. Too many new people unconnected with the pits had come to the village. One small aspect of this was his unwillingness to shop at the store any longer. The older shop assistants knew him and understood that, because of his damaged hand, they had to fold his fingers around his change. The newer ones did not know this and he felt embarrassed about it, so he gave up shopping.

The life review was not all rose-coloured. Throckley had changed considerably since the Second World War. The Throckley

pits themselves closed in 1953 and 1954. The 'Newcastle Journal'
headline on the closure of the Maria said: 'The Dying Day of a
Pit. New Jobs for the Men but they mourn ...' (13 March 1953).
My grandfather's comment was that it would be a better day
when all the bloody pits were closed. Although the Throckley
men transferred to other pits, Throckley lost its character as a
mining village. Work was available elsewhere and in different
industries, Throckley having by this time been completely
absorbed into the conurbation of industrial Tyneside. Entirely
new symbols of status and self-respect had replaced the old ones
of respectability and hard work. It would take several more
volumes to chart this argument but affluence, social mobility,
private housing, the welfare state and industrial diversification
had produced a subtle metamorphosis in social relations which all
of us alike have still to understand properly.

But in the course of it some things, by not fitting into the
increasingly idealised image of the past, were thought by my
grandfather to be lost. Community, neighbourliness, recognition,
sharing and contentment had all disappeared. Gloria articulated
this for me as she tried to sum up her grandfather. 'He formed
his own opinions,' she said.

> He knew he was an ordinary man, born to be a worker and
> didn't expect any different. He was more of a tradition man.
> He didn't like us to be greedy or envious. See to your needs
> first, then your wants. He used to say that to us.

And she amplified the point while talking about how he got his
coals. The coal allowance for retired miners was never quite
sufficient. My grandfather had supplemented this with coal picking, but later on he had often enough to buy coal from pitmen
who had had their free loads delivered. He could never bring
himself actually to pay the money over. He got Eva to do that.
He was disgusted that he should have to pay for coal. 'I've
never had to pay for coal in me life' is what he told Gloria,
contrasting the treatment of old people in his young day with
current attitudes. But generally he did not complain; he simply
acknowledged that times had changed, largely for the better.

Much of his picture of social change is, however, false, at
least in the sense that the economic changes which had brought
affluence in the 1950s and early 1960s were not of a type which
either gave ordinary people control of their lives or could per-
manently shift the north-east of England from its long-term
dependence on declining industry or arrest its pattern of long-
term industrial decline (see, for example, Benwell Community
Project 1979; Rowntree Research Unit, 1974).

The Second World War was a fulcrum of change, but the long-
term decline of staple industry in the region was not halted.
Such new industry as did arrive was in the field of semi-skilled
or unskilled work, often in units which were the branch factories
of larger companies, many of them multinational (Benwell Com-

munity Project, 1979; Austrin and Beynon, 1980). Capital from the nationalised coal industry found its way into finance, banking, insurance and property development. Ian Harford has shown how the fortunes of local capitalists in west Tyneside were recycled out of the area's older industry into property and multi-nationals (Benwell Community Project, 1979). In the case of the Throckley coal company, some of the capital at least, it seems, went into property, and some into mining operations in South Africa (Benwell Community Project, 1979). And with respect to a whole range of social and economic indicators, the north-east of England can be shown to have remained a relatively disadvantaged and depressed area (Taylor and Ayres, 1969).

Such a wide-ranging review, however, was not available to my grandfather. For him a feeling that throughout his life much had improved was rooted in experience and real enough.

On 10 February 1965, I received the following letter from my mother:

Just a short letter to let you know, your Granda died at 20 min to 2 this morning. Jack Brown and Sadie came to seek me at 1 a.m. He was conscious when I went in, he knew me. He asked for another pillow to be put under his head, then he took hold of my hand, held it until he died. He slept away peacefully. Uncle Jim Brown was there, too also Jack Brown, Sadie and Eva. Jim and I laid him out. He is being cremated, but when, I don't know.... I loved my Dad very much but I am happy he is with my mother. I saw he had changed when I was up on Mon night. He had had three falls this last week. He has died with a nasty black eye.

He had a large funeral despite the vile weather. Throughout his life he had managed to remain independent; seen in this light his death was a triumphant tilt at circumstance. Right to the end he kept that autonomy, dying where he had always preferred to be, in his own home with his family.

13 CONCLUSION

The life of one man can at best be but an illustration of the lives and experience of a whole social group. And since the conventions of society limit strictly which parts of ourselves, our thoughts and feelings, we can legitimately convey to others, it is inevitable that a biography, itself reflecting the selective interests of the biographer, runs the risk of distorting seriously the reality it seeks to describe. There must always be gaps in the record which no research technique can fill. Equally, the conventions of society - perhaps, too, the social and psychological needs of individuals - encourage selective portrayals of the self so that people can project and maintain, although not always successfully, a positive image of themselves in the eyes of others. The record or the data around which a biography must be written - irrespective of whether the data are written down or, as in this case, passed on through interviews and personal recollections - is necessarily partial and this must be taken into account in writing.

Two methods have been used in this book to control such distortions. The first is to check all accounts of the subject against other historical materials. This is the method of triangulation. It involves, at a minimum, organising the data in a strict chronology so that the personal records can be examined for their coherence and consistency against the logic of the situation they seek to portray and other kinds of data. To think in strict chronological terms is an historian's skill. Many of the respondents I spoke to found it difficult to locate their observations precisely in time. My job was to take them back through their experience to help them to recall the past more accurately. This method works. But it presupposes a great deal of background knowledge on the part of the interviewer. It involves more than one visit. In fact it is a process, not of interviewing, but of dialogue. This leads to the second method.

The account of the subject of the biography given here is not, in a sense, my own. The words are mine but the account itself was built up co-operatively. I have tested my interpretations out on others; we have discussed whether they 'ring true'. When none of us knew what my grandfather's thoughts or feelings on a particular issue were we attempted to imagine them, checking various possibilities against what we already knew of the man and of other men like him. The result is not the truth of the matter; rather it is a plausible account, something that to those who knew him well is consistent with what they really knew of

him. This process, of course, is endless and it is not without
its own special results. I think all of us have learned something
through this work, not just about the man we discussed, but
about ourselves.

The data generated in this way, for all the distortions they
necessarily contain, are vital for history. Despite the study's
limited focus on the life of one man and the changes which took
place in a small corner of a large coalfield, the larger generalisations, which both history and sociology seek to make about
change in society, must at least be consistent with the patterns
of change in social structure, in thought, feeling and relationships described here. The reason for this is a point about the
logic of generalisation itself.

James Brown is portrayed here in some ways as a symbol for
a whole group of men. I have not made him out to be representative of Northumberland miners of his generation; indeed, what
have interested me most are the ways in which he was unique.
But uniqueness can only be recognised through comparison.
The class position my grandfather occupied was that of a worker.
As a worker in the coal industry his experience of class was different from that of workers in other industries, e.g. agriculture
or shipbuilding. And he sensed himself to be different. But even
as a worker in coal his experience was different from that of
miners in other coalfields of Britain and certainly coalfields
abroad. The comparative study of mining is not so well developed
that we can state systematically what explains those differences.
From the work of Rimmlinger (1959) and Harrison (1978) and
Martin Bulmer (1975) and M. Daunton (1979a) a list of the factors
explaining such differences would certainly include the following:

1 Age of the coalfield: technical methods.
2 The geological character of the coalfield.
3 Structure of ownership of the industry.
4 Size of undertakings; company policies; character
  of the coal market served.
5 The degree of isolation of the community.
6 The structure of working relationships underground.
7 The pattern of housing tenure among miners.
8 Political and ideological environment of the labour
  movement.
9 Religion.
10 The structural position of the coal industry in relation
   to the rest of the economy.
11 The character and commitment of local trade union
   and political leaders.
12 The nature of government involvement in the industry.

These and other interdependent factors combine in different
ways in different pits and coalfields to define subtle differences
of class position and experience among different generations of
pitmen.

Within the same group of men there are further differences.
Some miners are or were active in politics and the union, others
were not. Some men were considered respectable, others not.
Some remained within the pits, others sought opportunities
elsewhere. It amounts to this: given the same task, digging coal
from the ground, men act differently. That does not render
generalisations about, for example, social classes invalid; rather
it makes them complex. In this approach, however, sociologists
and historians have a powerful tool to probe the rich experience
of different groups of men, and to see in that experience the
way in which the larger structures of politics, trade unionism
and class have been refracted through different levels of organisation from work and family life, through trade union lodges
to the labour movement as a whole, and through that to changes
in government and the state itself. That experience constitutes
a mosaic in H. Becker's sense (1971). My account must be seen
as a limited description of only one part of it.

Pursuing the argument that the attitudes and outlook of working men are valid in terms of their own experience, I have tried
to show in this book how the particular generation of miners
represented here, working for a relatively small, paternalistic
coal company in a village isolated and penetrated by Methodism,
developed attitudes which, while supportive of a tradition of
Labour (or, in the nineteenth century, Lib-Lab) politics, nevertheless, largely through relatively stable market conditions
peculiar to their pits, acquiesced in industrial attitudes which
were conciliatory and found class-based politics dangerous and
almost alien. Younger men, partly influenced by a more radical
ideological tradition, acted and thought differently, and even
within the group of older miners differences in the character of
union leadership in the Throckley lodges resulted in varying
patterns of industrial relations. In these respects Throckley was
different from other villages in the Northumberland coalfield. My
work cannot prove this proposition; but it at least opens up the
question so that what made Throckley unique can become part of
a more general understanding of the character of mining communities.

The community is described here as a *constructed community*.
The coal company sought to employ high quality, temperate
labour and to keep them. It wanted a community and not a labour
camp. The men who came to Throckley built a life for themselves,
as far as they could, free of the constraints of the company and
its rules and the vicissitudes of winning coal. The community
which developed had 'that necessary habit of mutuality' which
many writers have detected as central to working-class communities (Jackson, 1968, p. 166). But its structures were not
simply defensive or as isolated as this image suggests; the union
lodge, the Labour Party, the co-operative store had an offensive
rationale, too, displaying at different points in time changing
images of a better society, adjusting strategies for achieving
it according to the opportunities they were presented with.

Solidarity through political and industrial action rather than simply through neighbourly relationships was a central feature of the community. But it was not a homogeneous community free of division, nor did the idea of community carry the same meaning for everyone who lived in Throckley. My grandparents were of a generation of Throckley families which had experienced the insularity, the mutuality and the interdependence of a relatively small community dominated by the coal owners. Their children, however, grew to maturity in a village which, increasingly after the First World War, responding to several different trajectories of social change, was losing its character as a mining community, and in which relationships of paternalism were breaking down. I have tried to show that larger changes in the structure of British society transformed both mining and mining communities and have offered a tentative theory of social change as a consequence of crisis and power. Again, however, the point of comparison arises. Throckley meant different things to different generations; it meant different things to men and to women. Above all, it was never a static community, its structures frozen in time.

Even its history is not static; this, too, is a matter of interpretation, and that interpretation, taking the form of a life review, feeds how people understand themselves and their society. History, as Robert Colls (1977, p. 198) has rightly pointed out 'is not a sovereign quantity somehow trailing out behind us like a great winding-away into the distance. The past has been but it lives only as much as society is aware of it'.

It has been a particular awareness of the past which has shaped this study. My account of what was significant in the lives of my grandparents has been built up by my family itself. History has been viewed selectively but it is a selectivity revealing much of what is important to them as people and to Throckley as a community. The list of what has been left out is infinite. But it is what has been put in that matters. The First World War, the 1926 strike, family life and so on: these are the themes which are important to their collective image of the past and their attitude to them is a measure of how far British society has changed. For there is double image: on the one hand a nostalgic picture of family and community, on the other a picture of poverty and squalor which they would be glad to forget.

This ambiguity was a strong element of my grandfather's view of his past. In the main, however, despite the difficulties and the defeats his view of it was one of progress. His historical imagination extended back further than his own experience; it travelled far into rural Norfolk in the nineteenth century and, measured against his image of the past, the present was an improvement. But he was not a reflective man. Little in his education or his daily routines prompted him to be so. And in the end what he felt most content about was that he had maintained over a long life his own self-respect. That, for him, is what mattered most about class. It was a question of dignity and the recognition of others.

It is the recognition of others which is at the core of the idea of culture which has been used in this book. To understand people from Throckley - or anywhere else - it is essential to grasp the totality of their way of life and the meanings which attach to their actions and to the social institutions which are the framework of those actions.

BIBLIOGRAPHY

BOOKS AND ARTICLES

Abel-Smith, B. and Townsend, P. (1965), 'The Poor and the Poorest', London, School of Economics, Occasional Papers on Social Administration, no. 17.

Abrams, P. (1963), The Failure of Social Reform: 1918-1920, 'Past and Present', April.

Addison, P. (1977), 'The Road to 1945', London, Quartet.

Arnold, M. (1908), 'Reports on Elementary Schools 1852-1882', London, Wyman & Sons.

Ashby, M.K. (1961), 'Joseph Ashby of Tysoe 1859-1919', London, Cambridge University Press.

Austrin, T., and Beynon, H. (1980), 'Global Outpost: The Working Class Experience of Big Business in the North East of England 1964-1979', discussion document, Department of Sociology and Social Policy, University of Durham.

Bateman, J. (1971), 'The Great Landowners of Great Britain and Ireland' (first published 1876), Leicester University Press.

Becker, H. (1970), Notes on the Concept of Commitment, in 'Sociological Work, Method and Substance', Harmondsworth, Middx, Allen Lane, The Penguin Press.

Becker, Howard (1971), Life History and the Scientific Mosaic, in 'Sociological Work, Method and Substance', Harmondsworth, Middx, Allen Lane, The Penguin Press.

Bellamy, J.M., and Saville, J. (eds) (1976), 'Dictionary of Labour Biography', vol. III, London, Macmillan.

Bendix, R. (1956), 'Work and Authority in Industry', London, Chapman & Hall.

Benney, M. (1946), 'Charity Main: A Coalfield Chronicle', London, Allen & Unwin.

Benwell Community Project (CDP) (1979), 'The Making of a Ruling Class: Two Centuries of Capital Development on Tyneside', Newcastle upon Tyne, Benwell Community Project.

Berger, B.M. (1968), The Sociology of Leisure: Some Suggestions, in E.O. Smigel (ed.), 'Work and Leisure', New Haven, Conn., College and University Press.

Berger, J. (1969), 'A Fortunate Man', Harmondsworth, Middx, Penguin.

Berger, P., and Luckmann, T. (1966), 'The Social Construction of Reality', Harmondsworth, Middx, Allen Lane, The Penguin Press.

Blumer, H. (1939), 'An Appraisal of Thomas and Znaniecki's "The Polish in Europe and America"', Critiques of Research in the Social Sciences, no. 1, Social Sciences Research Council, New York.

Bourdieu, P. (1974), The School as a Conservative Force, in J. Eggleton (ed.), 'Contemporary Research in the Sociology of Education', London, Methuen.

Briggs, A. (1968), 'Victorian Cities', London, Penguin.

Bulmer, M. (1975), Sociological Models of the Mining Community, 'Sociological Review', no. 23.

—(1978), 'Mining and Social Change', London, Croom Helm.

Burke, P. (ed.) (1973), 'A New Kind of History from the Writings of Febvre',

London, Routledge & Kegan Paul.

Burnett, J. (1977), 'Useful Toil: Autobiographies of Working People from the 1820s to the 1920s', Harmondsworth, Middx, Penguin.

Burns, T. (1967), A Meaning in Everyday Life, 'New Society', 25 May.

Calder, A. (1969), 'The People's War', London, Panther.

Calvocoressi, P., and Wint, G. (1974), 'Total War', Harmondsworth, Middx, Penguin.

CDP (1977), 'The Costs of Industrial Change', London, CDP Inter-Project Editorial Team, Mary Ward House.

Chaplin, S. (1968), 'The Thin Seam', London, Pergamon Press.

—(1978), Durham Mining Villages, in M. Bulmer (ed.), 'Mining and Social Change', London, Croom Helm, 1978.

Cole, G.D.H. (1945), 'A Century of Co-operation', London, Co-operative Union.

Cole, M. (1977), The General Strike, 'Bulletin of the Society of Labour History', no. 34, Spring.

Colls, R. (1976), Oh Happy English Children! Coal, Class and Education in the North East, 'Past and Present', no. 75.

—(1977), 'The Collier's Rant. Song and Culture in the Industrial Village', London, Croom Helm.

Constantine, S. (1979), Amateur Gardening and Popular Recreation in the 19th and 20th Centuries, unpublished manuscript, Department of History, University of Lancaster.

Court, W.H.B. (1951), 'Coal', London, HMSO.

Daunton, M. (1979a), Down the Pit: Work in the Great Northern and South Wales Coalfields 1880-1914, mimeo, Durham University (forthcoming in 'Economic History Review').

—(1979b), Miners' Houses: The Great Northern and South Wales Coalfield, mimeo, Durham University (forthcoming in 'International Review of Social History').

Davidoff, L. (1974), Mastered for Life: Servant and Wife in Victorian and Edwardian England, 'Journal of Social History', vol. 7, no. 4.

Davies, S. (1963), 'North Country Bred: A Working Class Family Chronicle', London, Routledge & Kegan Paul.

Davison, J. (1973), 'Northumberland Miners' History 1919-1939', Newcastle upon Tyne, Newcastle Co-operative Press.

Dennis, N., Henriques, F., and Slaughter, F. (1956), 'Coal is Our Life', London, Eyre & Spottiswoode.

Dückershoff, E. (1899), 'How the English Workman Lives', London, P.S. King & Son.

Dyhouse, C. (1978), Working Class Mothers and Infant Mortality in England 1895-1914, 'Journal of Social History', vol. 12, no. 2, Winter, pp. 248-68.

Elias, N. (1974), Towards a Theory of Communities, in C. Bell (eds), 'The Sociology of Community: A Selection of Readings', London, Frank Cass.

Engels, F. (1968), 'The Condition of the Working Class in England' (first published 1844), London, Allen & Unwin.

Entwistle, H. (1978), 'Class, Culture and Education', London, Methuen.

Evans, George Ewart (1976), 'From Mouths of Men', London, Faber & Faber.

Faraday, A., and Plummer, K. (1979), Doing Life Histories, in 'Sociological Review', vol. 27, no. 4, November.

Farman, C. (1972), 'The General Strike, May 1926', London, Hart Davis.

Febvre, L. (1973), A New Kind of History, in Burke, 1973.

Fischer, D.H. (1976), The Braided Narrative: Substance and Form in Social History, in A. Fletcher (ed.), 'The Literature of Fact: Selected Papers from the English Institute', New York, Columbia University Press.

Flynn, M.W. (1976), Social Theory and the Industrial Revolution, in T. Burns and S.B. Saul (eds), 'Social Theory and Economic Change', London, Tavistock.

Fox, A. (1971), 'A Sociology of Work in Industry', London, Collier-Macmillan.

Frankenberg, R. (1966), 'Communities in Britain', Harmondsworth, Middx, Penguin.

Garside, W.R. (1971), 'The Durham Miners 1919-1960', London, Allen
& Unwin.
Gerth, H., and Mills, C. Wright (1954), 'Character of Social Structure: The
Psychology of Social Institutions', London, Routledge & Kegan Paul.
Goldthorpe, J. (1959), Technical Organisation as a Factor in Supervisor
Worker Conflict, 'British Journal of Sociology', vol. 10.
Goodfellow, D.M. (1941), 'Tyneside, The Social Facts', Newcastle upon Tyne,
Co-operative Printing Society.
Goodrich, C. (1975), 'The Frontier of Control' (first published 1920), London,
Theo Pluto Press.
Gregory, R. (1968), 'The Miners and British Politics 1906-1914', London,
Oxford University Press.
Hall, B.T. (1908), 'Working Men's Clubs: Why, and How to Establish and
Manage Them', London, Working Men's Club and Institute Union.
Hammond, J.L., and Hammond, B. (1920), 'The Skilled Worker 1760-1832',
London, Longmans.
Hargreaves, D. (1967), 'Social Relations in a Secondary School', London,
Routledge & Kegan Paul.
Harris, C.C., and Rosser, C. (1965), 'The Family and Social Change', London,
Routledge & Kegan Paul.
Harrison, R. (ed.) (1978), 'Independent Collier: The Coal Miner as an
Archetypal Proletarian Reconsidered', Hassocks, Sussex, Harvester.
Harvey, G. (c.1918), 'Capitalism in the Northern Coalfield', Pelaw-on-Tyne
(copy in Newcastle City Library).
Hayler, G. (1897), 'The Prohibition Movement', Newcastle, J. Dowling & Sons.
Heslop, Harold (1946), 'The Earth Beneath', London, T.V. Boardman.
'History Workshop: A Journal of Socialist Historians', Ruskin College, Oxford.
Hoggart, Richard (1957), 'The Uses of Literacy', Harmondsworth, Penguin.
Jackson, B. (1968), 'Working Class Community', Harmondsworth, Middx,
Penguin.
Johnson, R. (1976), Notes on the Schooling of the English Working Class
1780-1850, in R. Dale, B.R. Cosin, G.M. Esland and D.F. Swift (eds),
'Schooling and Capitalism: A Sociological Reader', Milton Keynes, The
Open University Press.
'Journal of the Club and Institute Union'.
Kendall, W. (1973), Labour Unrest Before the First World War, in
D. Rubenstein (ed.), 'People for the People', London, Ithaca Press.
Kerr, C., and Siegel, A. (1954), The Inter-Industry Propensity to Strike:
An International Comparison, in A. Kornhauser (ed.), 'Industrial Conflict',
New York, McGraw-Hill.
Kirby, M.W. (1977), 'The British Coalmining Industry, 1870-1946', London,
Macmillan.
Laquer, T.W. (1977), Working Class Demand and the Growth of English
Elementary Education 1750-1850, in L. Stone (ed.), 'Schooling and Society:
Studies in the History of Education', Baltimore, Johns Hopkins University
Press.
Lawson, J. (1949 edn), 'A Man's Life', London, Hodder & Stoughton (first
impression 1932).
McCord, N. (1979), 'North East England: The Region's Development 1760-1960'
London, Batsford.
McDonald, G. (1975), The Defeat of the General Strike, in Peele and Cook,
1975.
Macfarlane, A. (1977), History, Anthropology and the Study of Communities,
'Social History', vol. 2, no. 5.
Marshall, T.H. (1965), 'Social Policy', London, Hutchinson.
Marwick, A. (1970), 'Britain in the Century of Total War', Harmondsworth,
Middx, Penguin.
Mason, A. (1970), 'The General Strike in the North East', Hull, University
of Hull Publications.
Masterman, C.F.G. (1911), 'The Condition of England', London, Methuen.
Meacham, S. (1977), 'A Life Apart: The English Working Class 1890-1964',

London, Thames & Hudson.
Mess, H.A. (1928), 'Industrial Tyneside: A Social Survey', London, Benn.
Miliband, R. (1973), 'Parliamentary Socialism: A Study in the Politics of Labour', London, Merlin.
Mills, C. Wright (1970), 'The Sociological Imagination', Harmondsworth, Middx, Pelican.
Mink, L.O. (1970), History and Fiction as Modes of Comprehension, 'New Literary History', vol. 1, no. 3.
Moore, R. (1974), 'Pitman, Preachers and Politics', London, Cambridge University Press.
Newby, H. (1977), 'The Deferential Worker', Harmondsworth, Middx, Penguin.
Newson, J. (1936), 'Out of the Pit', Oxford, Blackwell.
Noel, G. (1976), 'The Great Lock-Out of 1926', London, Constable.
Norris, G.M. (1978), Industrial Paternalist Capitalism and Local Labour Markets, 'Sociology', vol. 12, no. 3.
Novak, T. (1978), Poverty and the State: A Study of Unemployment and Social Security in Britain, PhD thesis, University of Durham.
Oakley, F.A. (1977), 'Housework', London, Robertson.
Page-Arnot, R. (1949), 'The Miners', London, Allen & Unwin.
Parkinson, G. (1912), 'True Stories of Durham Pit Life', 3rd edn, London, Charles H. Kelley.
Peele, G., and Cook, C. (eds) (1975), 'The Politics of Reappraisal 1918-1935', London, Macmillan.
Pelling, H. (1968), 'Popular Politics and Society in Late Victorian Britain', London, Macmillan.
—(1970), 'Britain and the Second World War', London, Fontana.
Phillips, G.A. (1976), 'The General Strike: The Politics of Industrial Conflict', London, Weidenfeld & Nicolson.
Pilgrim Trust (1938), 'Men Without Work', London, Cambridge University Press.
Priestley, J.B. (1934), 'English Journey', London, Heinemann (1968 edn).
Purdue, A.W. (1974), Parliamentary Elections in the North East of England 1900-1906, Master of Letters thesis, University, Newcastle upon Tyne.
Redmayne, R.A.S. (1923), 'The British Coal-Mining Industry During the War', Oxford, Clarendon Press.
Renshaw, P. (1975), 'The General Strike', London, Eyre Methuen.
Rex, J. (1961), 'Key Problems in Sociological Theory', London, Routledge & Kegan Paul.
Rickman, H.P. (1976), 'Dilthey, Selected Writings', London, Cambridge University Press.
Rimmlinger, G.V. (1959), International Differences in the Strike Propensity of Coal Miners: Experience in Four Countries, 'Industrial and Labour Relations Review', vol. 12, no. 3.
Roberts, R. (1971), 'The Classic Slum', Harmondsworth, Middx, Penguin.
—(1976), 'A Ragged Childhood', London, Fontana.
Rosser, C. and Harris, C.C. (1965), 'The Family in Social Change', London, Routledge & Kegan Paul.
Rowe, D.J. (1973), 'The Economy of the North East in the Nineteenth Century', North of England Open Air Museum Beamish, County Durham.
Rowe, J.F.W. (1923), 'Wages in the Coal Industry', London, P.S. King & Sons.
Rowntree Research Unit (1974), Aspect of Contradiction in Regional Policy: The Case of North East England, 'Regional Studies', vol. 8.
Schwartz, H., and Jacobs, J. (1979), 'Qualitative Sociology: A Method to the Madness', London, Collier-Macmillan.
Seeley, J.Y.E. (1973), Coal Mining Villages of Northumberland and Durham: A Study of Sanitary Conditions and Social Facilities 1870-80, MA thesis, Newcastle University.
Sennet, R., and Cobb, J. (1972), 'The Hidden Injuries of Class', Cambridge, Cambridge University Press.

Simpson, J.B. (1900), 'Capital and Labour Employed in Coal Mining',
  Newcastle upon Tyne, R. Robinson copy in Newcastle Central Library.
Skipsey, Joe (1976), 'Joe Skipsey, Pitman Poet of Percy Main (1832-1903)',
  Metropolitan Borough of North Tyneside, Preston Grange County Primary
  School.
Stacey, M. (1969), The Myth of Community Studies, 'British Journal of
  Sociology', vol. XX, no. 2.
Stearns, P.N. (1972), Working Class Women in Britain, 1890-1914, in
  M. Vincinus (ed.), 'Suffer and be still: Women in the Victorian Age',
  Bloomington, Indiana University Press.
Stevenson, J. (1977), 'Social Conditions in Britain Between the Wars',
  Harmondsworth, Middx, Penguin.
Stevenson, J., and Cook, C. (1977), 'The Slump: Society and Politics
  During the Depression', London, Cape.
Straker, W. (1916), An Outline of the Wages Question, in 'Minutes of the
  Northumberland Miners' Association', July.
Taylor, G., and Ayres, N. (1969), 'Born and Bred Unequal', London,
  Longmans.
Taylor, J. (1973), 'From Self Help to Glamour: The Working Man's Club
  1860-1972', Ruskin College Oxford, History Workshop Pamphlet.
Thomas, W.I., and Znaniecki, F. (1958), 'The Polish Peasant in Europe and
  America', vol. 11, New York, Dover.
Thompson, E.P. (1968), 'The Making of the English Working Class',
  Harmondsworth, Middx, Penguin.
—(1977), 'William Morris: Romantic to Revolutionary', London, Merlin.
Thompson, P. (1977), 'The Edwardians: The Remaking of British Society',
  London, Paladin.
—(1978), 'The Voice of the Past: Oral History', Oxford, Oxford University
  Press.
Thorpe, E. (1970), Coal Port: An Interpretation of 'Community' in a Mining
  Town, Durham, University of Durham Rowntree Research Unit.
Titmuss, R. (1950), 'Problems of Social Policy', London, HMSO.
Tobias, J.J. (1967), 'Crime and Industrial Society in the Nineteenth
  Century', Harmondsworth, Penguin.
Tremlett, G. (1962), 'The First Century', London, The Working Men's Club
  and Institute Union Ltd.
Trist, E.L., Higgin, G., Murray, H., and Pollock, A.B. (1963), 'Organisa-
  tional Choice', London, Tavistock.
Wade, E. (1978), The Putter of the Northumberland and Durham Coalfield,
  'History Bulletin', no. 12, October, North East Group for the Study of
  Labour.
Welford, T., 'Men of Mark 'Twixt Tyne and Tweed', London, Walter Scott.
Williams, R. (1977), The Social Significance of 1926, 'LLAFUR', the Journal
  of the Society for the Study of Welsh Labour History, vol. 2, no. 2, Spring.
Williams, W.H. (1937), 'Coal Combines in Northumberland', Labour Research
  Department, London, Transport House.
Williamson, Bill (1973), The Leek, in Karl Miller (ed.), 'A Second Listener
  Anthology', London, British Broadcasting Corporation.
—(1976), Shaping the Sociology of Leisure, 'Journal of Psycho-Social
  Aspects', Occasional Paper no. 2, April.
Williamson, W. (1980), Class, Culture and Community: A Study of Social
  Change through Biography, PhD thesis, University of Durham.
Willis, P. (1978), 'Learning to Labour. How Working Class Kids Get Working
  Class Jobs', London, Saxon House.

GOVERNMENT REPORTS, COMMISSIONS, ETC.

Blakiston, J.R. (1886), 'General Report of the Year', Board of Education
  Reports.
Coleman, Mr (1882), 'Report on Northumberland', Reports of the Assistant

Commissioner, Royal Commission on Agriculture, London, HMSO.
Commission of Inquiry into Industrial Unrest (1917), 'Report of the Com-
   missioners for the North Eastern Area (No. 1 Division)', London, HMSO.
Hammond, J.C. (1867-8), 'Northumberland Reports from Commissioners',
   Schools Inquiry Commission, vol. XXVIII, part VII.
Lambert, Sir A. (1945), A Farewell Message from the Regional Commissioner
   to the Personnel of the Northern (No. 1) Region, Newcastle Central Library.
Mines Inspectorate (annual), 'List of Mines', London, HMSO.
Ministry of Labour (1934), 'Report of Investigations into the Industrial
   Conditions in Certain Depressed Areas', London, HMSO, Cmnd 4728.
'Report from the Select Committee on Mines: Minutes of Evidence', 1866.
'Report of H.M. Inspector of Mines (Northern Division (No. 2))' (1914),
   London, HMSO.
'Royal Commission on the Poor Laws and Relief of Distress' (1909), Appendix
   no. CXVI, vol. V.
Technical Advisory Committee (1945), 'Coal Mining' (Reid Report), London,
   Ministry of Fuel and Power, Cmnd 6610.

NEWSPAPERS

'Blaydon Courier' (Gateshead Public Library).
'Hexham Courant' (Newspaper Office, Hexham).
'Newcastle Journal' (Newcastle Central Library).
'Newcastle Weekly Chronicle' (Newcastle Central Library).

LOCAL GOVERNMENT DOCUMENTS

Chief Constable of Northumberland, File: Official Circulars re Emergency,
   Northumberland Record Office, NRO NC/1/20/(1926).
Chief Constable of Northumberland, Standing Joint Committee Minutes,
   NRO CC/CM/SJ.
Medical Officer of Health (MOH), Newburn UDC, Annual Reports,
   Tyne-Wear Archive, West Denton Library.
Newburn UDC, Housing Committee Minutes, Tyne-Wear Archive.
Northumberland Records Office, tape-recorded interviews with Mrs Hall of
   Heddon (b. 1894) (T/114) and with Miss Sarah Elliot (T/117).

COAL COMPANY, TRADE UNION AND EMPLOYERS' ASSOCIATION
DOCUMENTS, ETC.

Dawson Papers, Northumberland Record Office, (NRO) 527/B/1-12.
Flynn, C.R. (1926), The General Strike: An Account of the Proceedings
   of the Northumberland and Durham General Council Strike Committee,
   Gateshead Public Library, Reference Section.
Lambert, Sir Arthur W. (1945), A Farewell Message from the Regional
   Commissioner to the Personnel of the Northern (No. 1) Region, Newcastle
   upon Tyne, copy in Newcastle Central Library.
Newburn and District Local Labour Party, Minutes, NRO 527/B/1.
Newburn and District Trades Union Council of Action, Minutes, 1926 (filed
   with the Dawson Papers), NRO 527/B/12.
Newcastle Trades Council of Action, 'Worker's Chronicle', no. 11, 1926.
Northumberland Coal Owners' Association, Minutes, NRO DL/C.
Northumberland Coal Owners' Association, Mutual Protection Association,
   Annual Report, NRO.
Northumberland Coal Owners' Association, Statistical Information,
   NRO NC8/C/1.
Northumberland Miners' Association (NMA), File on Output Committees,
   1920, Burt Hall, Newcastle upon Tyne.

Northumberland Miners' Association, File on Replies to Questionnaires,
   1924-33, Burt Hall, Newcastle upon Tyne.
Northumberland Miners' Association, Minutes (of Executive and Council).
   NRO 759/68.
Spen and District Trades and Labour Council, 'Strike Bulletin', no. 1,
   1926, Gateshead Public Library.
Steam Collieries Defence Assocation, Minutes, Northumberland Record Office.
Throckley Coal Company, Minutes, NRO/407 (Throckley Colliery Records).
Throckley Federation of Miners, Aged Miners' Coal Fund, Minutes, Tyne-Wear
   Archive, 1059/1.
Throckley Isabella Miners' Lodge, Minutes, to be deposited in Northumberland
   Head Office.
Throckley Miners' Welfare Fund, Correspondence File, Burt Hall, Newcastle
   upon Tyne.
Trades Union Congress, 'British Worker', no. 1, 11 May 1926, no. 4,
   Gateshead Public Library.
Wansbeck Divisional Labour Party, Minutes, NRO 527/B/1.

SCHOOL RECORDS

Belmont School Log Books, Durham County Record Office.
Heddon-on-the-Wall School Log Books, School Office, Heddon-on-the-Wall.
Newburn Church Magazines, The Vicarage, Newburn.
Throckley School Log Book, Northumberland Record Office.
Throckley Wesleyan Sunday School Minutes, Tyne-Wear Archive, 1093/13.

MISCELLANEOUS

Aged Miners' Coal Fund, Minute Book, Tyne-Wear Archive, 1059/1.
General Strike Pamphlets, Gateshead Public Library.
Northumberland Miners' Association, Correspondence re Welfare Fund,
   Burt Hall.
Stephenson Family Scrap Book (photocopies from West Denton Public Library).
Throckley District Co-operative Society, Balance Sheets, Tyne-Wear Archive.
Throckley Union Jack Club, Minutes of Committee, Throckley.

INDEX